Enterprise Guide to Gaining Business Value from Mobile Technologies

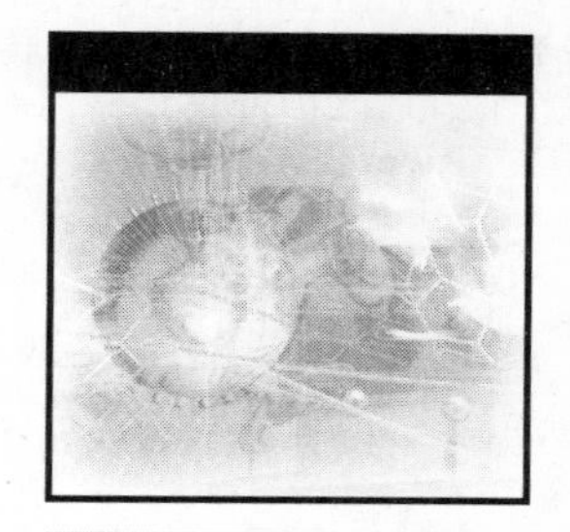

Enterprise Guide to Gaining Business Value from Mobile Technologies

Adam Kornak
Jorn Teutloff
Michael Welin-Berger

Wiley Publishing, Inc.

Enterprise Guide to Gaining Business Value from Mobile Technologies
Published by:
Wiley Publishing, Inc.
111 River Street
Hoboken, NJ 07030-5774
www.wiley.com/compbooks

Published simultaneously in Canada

Library of Congress Cataloging-in-Publication Data is available from the publisher.

ISBN: 0-471-23762-0

Manufactured in the United States of America

10 9 8 7 6 5 4 3 2 1

Credits
Vice President and Publisher: Joseph B. Wikert
Executive Editor: Carol A. Long
Editorial Manager: Mary Beth Wakefield
Development Editor: Kezia Endsley
Production Editor: Felicia Robinson
Text Design & Composition: Wiley Composition Services

Contents

Acknowledgments

There are a number of people we would like to thank, all of whom were instrumental in bringing this book to our readers.

Microsoft

Microsoft's Mobile Devices Division

The Microsoft EMSS team

The Legal team at Microsoft

The Scotia Bank team (and Scotia Bank) for allowing us to use the case study

The Microsoft Financial Services Group and the VSU

Cap Gemini Ernst & Young

John Distefano

Brent Whitman and the Legal team

Marv Whitson and the Mobility Team

Hans Torin

Last but not least, we'd like to thank our families for their continued support during the many evenings and weekends that it took to write this book.

About the Authors

Adam Kornak (Chicago, Illinois) is currently employed by Microsoft Corporation and he specializes in mobility solutions and strategy for the financial services industry. He has over 14 years of experience in the field of information technology and systems design. Adam has previously written a number of books and articles and is the lead author of CGE&Y's *Guide to Wireless Enterprise Application Architecture,* published by John Wiley & Sons. You can reach Adam at akornak@hotmail.com.

Jorn Teutloff (Los Angeles, California) is President of JV-Global, LLC, a professional services firm focusing on strategic business consulting, workforce procurement, and outsourcing solutions. For over 8 years, he has led project teams in the areas of strategic planning, business conceptualization, and business-process reengineering at clients in various industries. Jorn is a coauthor of CGE&Y's *Guide to Wireless Enterprise Application Architecture.* You can reach Jorn at jteutloff@jvglobal.com.

Michael Welin-Berger (Stockholm, Sweden) is mobile business director at Cap Gemini Ernst & Young. With 4 years of experience with mobile solutions and 12 years working with infrastructure and the Internet before that, Michael Welin-Berger has extensive experience in demonstrating how to benefit from new technologies. Michael is currently sharing his time between Europe and the United States, helping large clients to build their roadmap when implementing mobile solutions and gaining the potential benefits. You can reach Michael at michael.welin-berger@telia.com.

Foreword

Much has changed in the mobile technologies landscape over the last couple of years, on both the vendor and user sides. The number of mobile and wireless product and service providers has grown dramatically, offering an ever-increasing array of mobile and wireless products and services. New breeds of smart wireless devices are serving as catalysts for a massive transformation of the wireless industry. These new devices will change the way phones are used, from simple voice and text messaging devices, to true information appliances that retain the convenience and ease-of-use of a mobile phone while having the power and flexibility of a personal computer. Consequently, software companies and mobile operators alike are racing to offer services and applications that "light up" the new breed of devices and to help fill businesses' and end-users' ever increasing desire for on-the-go information. Within a few years, today's $500 billion mobile phone industry will transition into an even larger industry focused on delighting end-users and businesses alike with a vast array of exciting and productivity-enhancing applications and services.

Over the last months, I have witnessed a growing number of enterprises accelerating the pace at which they experiment with and come to adopt mobile solutions in their daily operations, both to serve line-of-business needs (such as field force automation and asset tracking) and to serve horizontal needs (such as e-mail). Leading-edge, or rather bleeding-edge, consumers too are experimenting with the services that have already been launched and are fueling the demand for new services and mobile applications. Especially in the United States, where the penetration of mobile and wireless has traditionally lagged behind the success in Europe and Asia, it is refreshing to see an ever-growing interest in this exciting technology. Within 3–4 years, Europe, Asia and the United States will no doubt be on par.

Whereas we previously saw only a few trailblazing companies slowly dipping their toes in the cold and unfamiliar waters of the mobile and wireless realms, today more and more organizations are beginning to realize that mobile technologies are here to stay. I am meeting more and more CIOs and business decision makers for whom it is not a matter of *if*, but *when*, they roll out mobile devices, services, and applications broadly within their organizations. Building on this understanding, today's enterprises are increasingly inquiring about the business value these technologies can provide. After all, one of the lessons learned from the dot-com crash is that no matter how sexy a technology might appear, implementing solutions without a sound footing in clearly understood business requirements has too often proved a recipe for overinvestment and sometimes even failure. This time around, organizations have wised up to the fact that there must be clear value propositions and business justification defined before significant efforts—human resources, time and money—are expended in conjunction with technology deployments.

The questions, then, on the minds of today's business leaders center more strongly than ever on the reasons for investing in technology. And this time around the focus is not on functionality per se, but the business fundamentals that drive technology deployment. Why should we be developing a mobile technology strategy? What are the business requirements that drive potential mobility solutions? How will those solutions affect our internal and external constituencies? And, most important, what are the strategic, qualitative, and quantitative value propositions that mobile technology can offer us?

This book, written by an experienced team of practitioners from the United States and Europe who have led multiple engagements in the mobility space, answer these questions with unique insight. They offer practical approaches that the audience can follow in the comfort of their own environments. This book is a valuable guide and inspiration for an audience that represents the business organization, not just the IT department. As we witness how mobile and wireless technologies continue to make inroads into today's organizations, there seems to be a tremendous need to learn about mobility, its business drivers, and how the technology can provide real value to those companies that are willing to deploy the technology.

The *Enterprise Guide to Gaining Business Value from Mobile Technologies* seeks to inform, educate, and inspire its audience. The book's sections guide the reader through the maze of this rapidly emerging industry, starting with a backgrounder on the terminology, drivers, devices, and networks, and then introducing a Value Web strategy analysis tool, presenting examples of mobile applications, and finally offering a hands-on guide to developing a mobile technology strategy at the reader's organization. Whether you are a business leader, an IT professional, or someone involved in the mobile industry at a mobile operator, device manufacturer, or solutions provider, I have no doubt you will find this book a valuable tool for crafting your own mobile technology strategy and a platform from which to launch the initiatives that will expand the reach of your organization.

—Juha Christensen

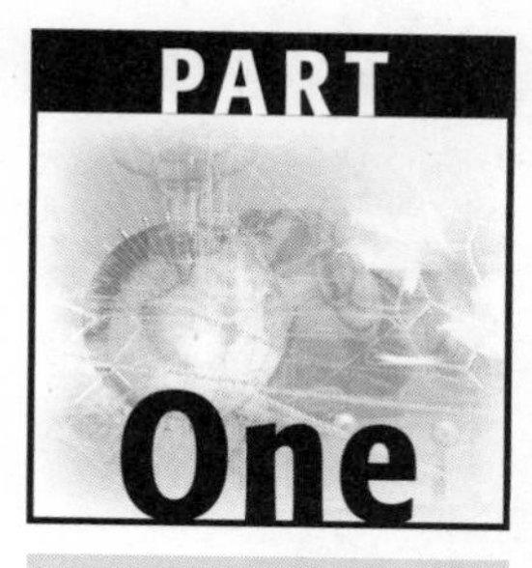

Introduction to Mobile and Wireless Technologies

CHAPTER

1

Wireless and Mobility Defined

Adam Kornak

Introduction

The terms *wireless* and *mobility* are quite often confused in the world of mobility. In fact, many mobile technology users consider these two terms as having the same or very similar definitions. However, the two words have very unique meanings, yet are used interchangeably, especially in the mobile business world. What then is the difference? *The American Heritage Dictionary* definitions are:

Wireless: "Having no wire or wires."

Mobility: "1. Capable of moving or of being moved readily. 2. Changing quickly from one condition to another."

(The American Heritage Dictionary, 3rd Edition, Delta Trade Paperbacks by Dell Publishing, copyright 1992)

As mentioned, in the world of mobile technology, the two terms wireless and mobility are often used interchangeably, especially when we talk about business applications for the enterprise. In any case, these definitions can get somewhat confusing, so let's talk about what they mean in the mobile business-applications sense. Hopefully, our definitions will give you a better idea of how you can apply wireless technologies and mobile technologies to your unique situation.

What Is Mobility?

There is no doubt that mobility can mean many things, depending upon the situation. Let's look at some interesting examples of mobility. Start by imagining yourself on a trip to your favorite vacation spot in Costa Rica. Generally, a flight in an airplane is required to get there. You are mobile in the sense that your physical person can be moved through various means to get to your destination. The device(s) or vehicles that you use to get to your destination are an airplane and probably a car or taxicab to drive you to your hotel while in sunny Costa Rica. The taxicab, the car, an airplane, or bicycle, are all examples of mobile devices or mobile objects. The hotel you stay in while in Costa Rica and the restaurant where you eat dinner are certainly not high on the mobility list. Besides yourself, the airplane, and the car, what other mobile devices might you use during your trip? Well, you may have taken along your calculator for expense management. You may have also used your personal digital assistant (PDA) to manage your vacation schedule while in Costa Rica. For that matter, assume that your son loves to play *Tetris* on your IPAQ device as well. These are all examples of mobile applications in the traditional sense. Let's take a look at mobile applications with more of a business perspective. As our dictionary definition explained, mobility is the capability of being moved readily. There's certainly much more to that definition that meets the eye, which is where we'd like to pick up. Our definition of mobility is:

> *The application of mobile devices and wireless technology to enable communication, information access, and business transactions from any device, from anyone, from anywhere, at anytime.*

You may have noticed that this definition includes the word *wireless*. The key pieces of this definition here are that mobility and wireless can go together and are not necessarily separate entities. For an additional perspective, see Figure 1.1.

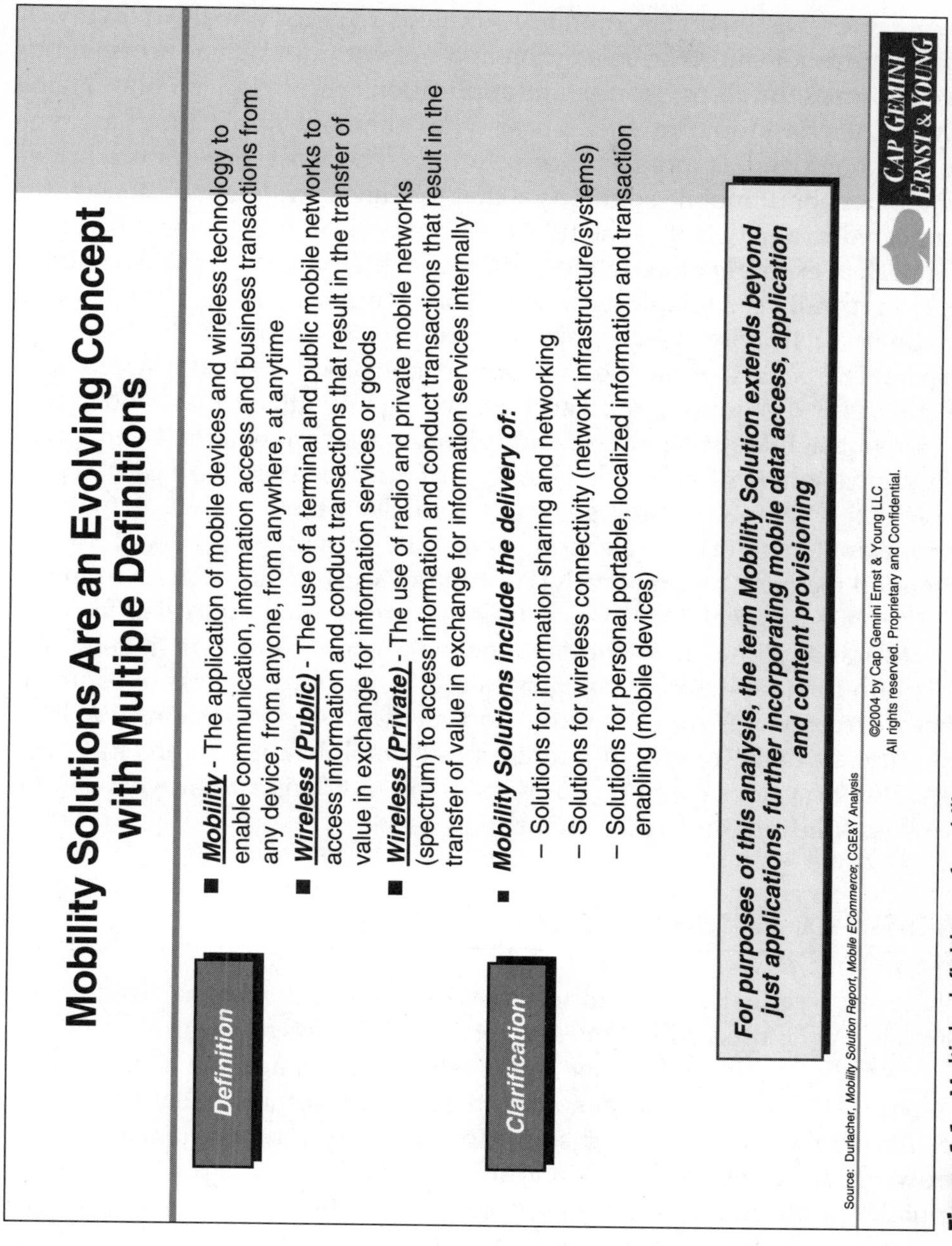

Figure 1.1 Multiple definitions of mobility.

As Figure 1.1 illustrates, mobility solutions may include the delivery of applications for information sharing and networking such as accessibility to corporate e-mail or groupware applications through a mobile device instead of a desktop computer. These types of mobile applications are typically categorized as mobile office solutions. The most important point to remember about mobile solutions is that the device need not be constantly connected to a network or the Internet to provide value. The mobile office is a perfect example of a mobile application. Many of us in the consulting world typically carry laptops or some sort of mobile device. Probably one of the most heavily used applications on the desktop or the laptop is e-mail. The beauty of most e-mail applications today is that a wired connection is not required. You can type the e-mail, attach files, and create documents while sitting in a car, on a plane, almost anywhere! When a mail message is sent in disconnected mode, it typically is stored in a cached outbox or other temporary storage mechanism. Once the user logs on to his or her normal network connection, voilá, the e-mail is delivered to the intended recipient automatically! As you can see, e-mail is a great example of a mobile application. You can manage your e-mail and calendar anywhere and anytime, even without a network connection. E-mail generally falls into the mobile office category, along with word processing, spreadsheets, groupware, Web browsing, and many other applications. We'll go into greater detail on the different types of mobile applications in Chapter 2. For now, let's continue our discussion and define what we mean by wireless solutions and applications.

What Are Wireless Applications?

There are really two categories of wireless solutions that need to be defined. As Figure 1.1 illustrates, one type of wireless application is a wireless "public" solution. In other words, it's the use of a terminal and public networks to conduct transactions and access information that ultimately results in the transaction of some sort of service or exchange. A Web browser such as Internet Explorer for Pocket PC is an example of a wireless "public" application. Unless you're browsing cached pages, a Web browser typically is of little use for obtaining up-to-the-minute stock quotes unless you're connected to the Internet. The same holds true for the wireless Web browser application that resides on a mobile device such as a WAP (Wireless Application Protocol) phone or a Pocket PC device. Other examples of wireless "public" applications are described in the following sections.

Real-Time Quoting Engines

These are applications that provide real-time information to the end-user typically for informational or analysis purposes. The key term to remember here is *real-time*, because most truly Wireless services supply essential information as it happens, such as emergency alerts, or a stock price drop. Typically, real-time information is required to make immediate and informed business decisions. Nonessential data, such as news articles, can usually be downloaded to a mobile device and read while on your train ride home or at home.

Location-Based Services

Location-based applications require the device to be constantly connected, or always on. It involves a two-way communication of information based on the location of a device (for promotions, emergency services, and so on). For example, the local hardware store can push coupons to a user's mobile phone as they're driving by the store. Location-based services have received some pushback from the consumers for security reasons. After all, who wants Big Brother watching over your shoulder, sending you messages at his whim? On the other hand, location services can also be used to locate individuals in need of help, as is done by E911 or enhanced 911 services. We'll cover these and other location services in more detail in the following section.

Wireless Portal/Commerce

Web portals such as Yahoo.com, MSN.com, and so forth require a live Internet connection to provide the door to enter their various services. The wireless Internet ultimately acts as that door. The Web portal can also be a corporate intranet portal that allows the employees of the company access address books or administrative applications such as ERP reports. The wireless device is another vehicle with which to access the portal. In the book *CGE&Y Guide to Wireless Enterprise Application Architecture*, we cover the various concepts behind wireless portals, including exchanges for B2B (business to business), B2C (business to consumer), B2E (business to employee), and so forth. For now, it's important to understand that wireless technology is a springboard for bringing the portal to life.

Sales Force Automation (SFA)

In today's world, the sales force plays an integral role not only in marketing a company's products to its clients, but also in building the strong business relationships that will ultimately lead to that "big deal." The technology that combines the power of a virtual sales team with up-to-the-minute decision-making processes is known as *sales force automation* (SFA). SFA is arguably a great deal more than just a client relationship management tool, but for our definition, it's a fair start. Generally, SFA applications fall into a far broader category known as customer relationship management (CRM). Sales teams are generally known as a mobile workforce, which makes them perfect candidates for leveraging mobile applications. SFA mobile applications can be as simple as a Personal Information Management (PIM) tool that manages schedules, to a sophisticated data warehouse generating monthly sales reports. We'll talk about what some of those SFA applications are in more detail in Chapter 9.

Global Positioning Systems (GPS)

One of the more captivating mobile technologies that has been around for quite some time is the global positioning system. The government has leveraged this technology for defense systems since the 1970s. More recently, the consumer and enterprise business markets are becoming aware of this technology; it's now being built for the everyday mobile user. A simple explanation of a global positioning system is that it provides the ability to pinpoint the exact location (within 10–50 feet) of an object carrying a GPS locator device, using satellites in outer space. An example of a GPS device is a golf ball tracking mechanism at your local golf course. Usually residing on a golf cart, a small television screen maps out the distance of a golf ball location to the final destination, the pin. It's designed to tell the golfer the proper golf club to use in any particular situation. By the way, the golf club also has the opportunity to provide additional services to the golfer through food or golf equipment ads. When golfers reach hole 9, they receive a message asking them if they would like to order food or beverages before reaching hole 10. The food and beverages are ordered and (theoretically) waiting for them once they complete the front 9.

Telematics

Last, but not least, is the technology that combines the best of all the worlds of mobility, *Telematics*. In one sense, Telematics combines location-based services, wireless portals, CRM, and GPS. Telematics is the use of wireless applications in vehicles, such as cars, buses, trains, and so on. Telematics applications typically provide a wireless connection to some service provider that allows the exchange of data between the vehicle and the service provider, like an automotive dealership. A real example of Telematics is General Motors' On-Star system. With On-Star, data about your automobile, such as when it's time for an oil change, is tracked and sent to the nearest dealership in your vehicle's area. You, as the vehicle's owner, are notified to complete the service. Other services provided include emergency services, stolen vehicle tracking, voice-activated wireless telephony, roadside assistance, and accident assistance to name a few. Telematics services are growing exponentially, as Figure 1.2 illustrates. We'll dig deeper into more Telematics examples in Chapter 9.

Worldwide Telematics Revenues

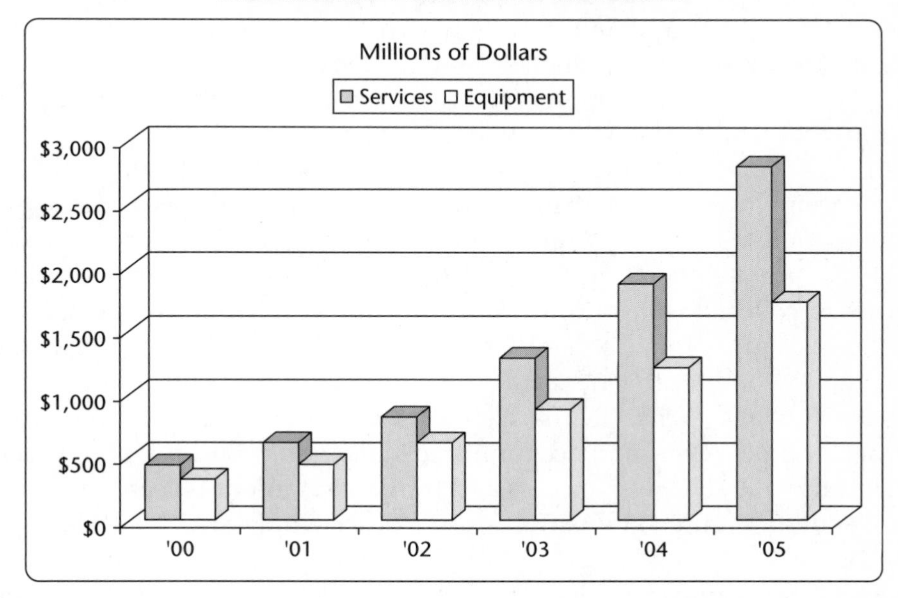

	'00	'01	'02	'03	'04	'05
Services	$450	$610	$832	$1,286	$1,887	$2,810
Equipment	$320	$460	$626	$872	$1,219	$1,730
Total	$770	$1,070	$1,458	$2,158	$3,106	$4,540

Source: Micrologic Research

Figure 1.2 The growth of Telematics services.

The other category of wireless applications that we need to discuss is wireless "private" applications. Referring back to Figure 1.1, you can see that our definition of wireless "private" applications involves the use of radio and private mobile networks (spectrum) to access information and conduct transactions that result in the transfer of value in exchange for information services internally. Now, let's decipher that quickly. The wireless "private" application has actually been used in the enterprise for many, many years. A good example is renting a car. Many automobile renting services, such as Avis, have been using ruggedized wireless devices to check cars back into their lots when a customer returns a rental car. Most of us don't even give it a second thought, because the service is very much behind the scenes.

When you drop off the car, the attendant supplies you with a receipt of payment literally within minutes of doing so. That attendant holds a commercial mobile device that is designed to withstand outside temperatures and can take being dropped (called a *ruggedized mobile device*). This device connects to an internal wireless radio network that accepts the transaction, looks up the renter's information as to when the car was dropped off, rates, and so forth. If the information is accepted, the device spits out a receipt of the transaction and off you go to your flight or next destination.

Other examples of wireless "private" networks include supply chain optimization for manufacturing processes, commercial Telematics, and wireless home networks. Wireless "private" applications generally exclusively use private spectrums, but in some instances can use the wireless Internet to conduct transactions. A wireless B2B exchange is an example. Usually, exchanges require many organizations to become members of the exchange to conduct business transactions. To reduce costs and provide easy access to the exchange, the Web is used as an access tool. Mind you, the back-end processes and accounting usually still occur in a private setting, but the initial browser access and transaction initiation can occur in a public setting. Figure 1.3 illustrates some other wireless and mobile solutions that we've applied to the utilities industry.

Without going into great detail on each of the solutions in Figure 1.3, you can see the enormous potential for leveraging mobility in the enterprise. The bottom line is, if a workforce is mobile, there likely exists mobile or wireless applications that improve processes and reduce costs. Hopefully, some of the topics covered in this book will help you understand the importance of mobility, where it is, and where it's going, and ultimately help you make intelligent decisions to apply mobile services that will affect your bottom line.

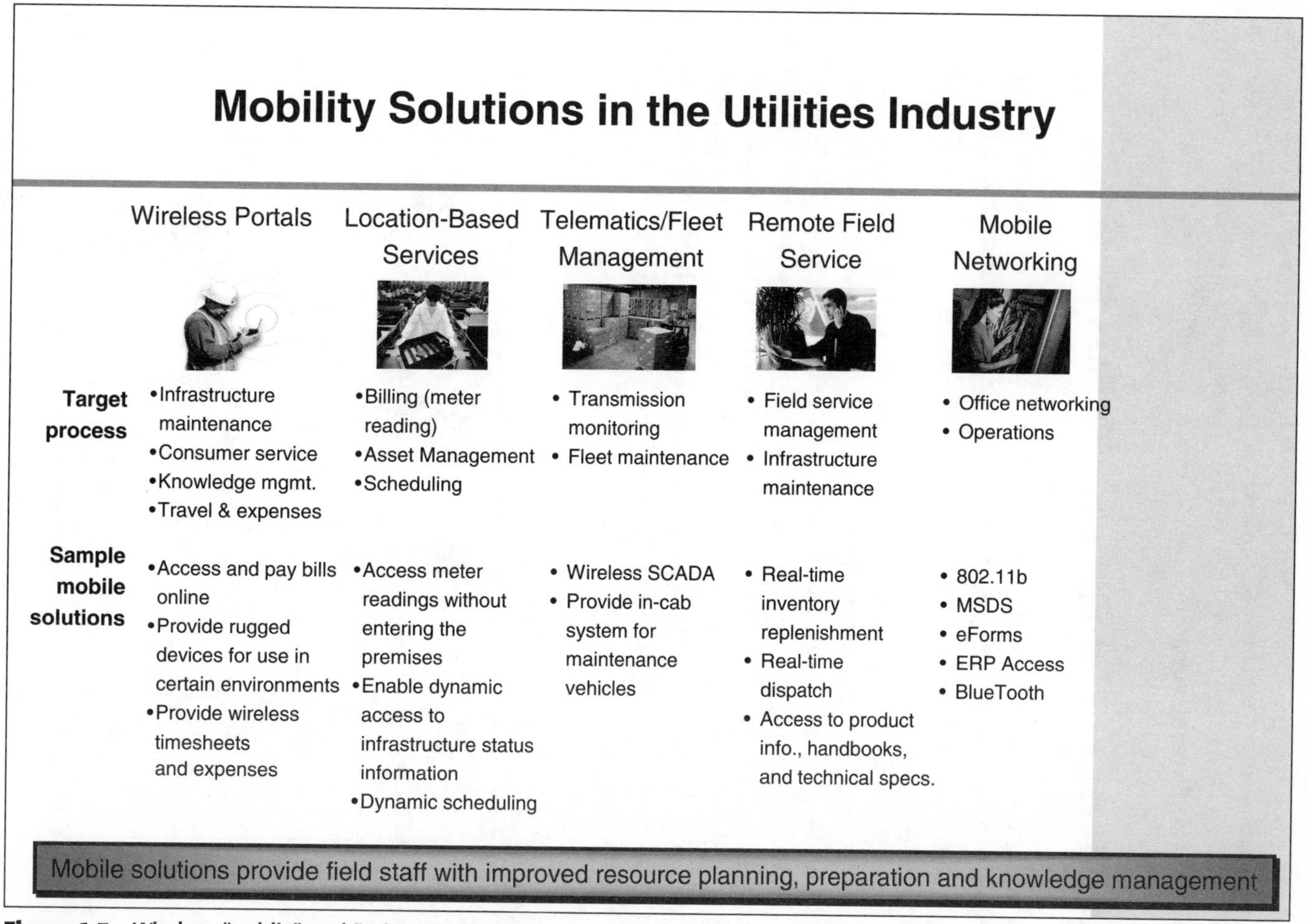

Mobility Solutions in the Utilities Industry

	Wireless Portals	Location-Based Services	Telematics/Fleet Management	Remote Field Service	Mobile Networking
Target process	•Infrastructure maintenance •Consumer service •Knowledge mgmt. •Travel & expenses	•Billing (meter reading) •Asset Management •Scheduling	• Transmission monitoring • Fleet maintenance	• Field service management • Infrastructure maintenance	• Office networking • Operations
Sample mobile solutions	•Access and pay bills online •Provide rugged devices for use in certain environments •Provide wireless timesheets and expenses	•Access meter readings without entering the premises •Enable dynamic access to infrastructure status information •Dynamic scheduling	• Wireless SCADA • Provide in-cab system for maintenance vehicles	• Real-time inventory replenishment • Real-time dispatch • Access to product info., handbooks, and technical specs.	• 802.11b • MSDS • eForms • ERP Access • BlueTooth

Mobile solutions provide field staff with improved resource planning, preparation and knowledge management

Figure 1.3 Wireless "public" and "private" solutions.

Now that we've covered the basic differences between what wireless applications and services are and what mobile applications and services are, let's go into more detail about the attributes and applications of mobility and wireless services.

Attributes of Mobility and Wireless Services

Now that we've talked about the differences and complements between a wireless and a mobile application, let's get a better understanding of the characteristics of the two and how they might be applied to a sample business scenario.

One of the first important points to mention regarding the attributes of mobility and wireless is the plethora of devices used to enable these applications. At last count, the number of mobile devices available to the marketplace was over 100 and rising. Many of these devices perform similar functions, whereas others are in a class of their own. We'll get into the various mobile devices available today and their characteristics in Chapter 3. For now, take a look at Figure 1.4, which gives you a high-level idea of the categories of devices and whether they fit into the wireless or mobile attribute category.

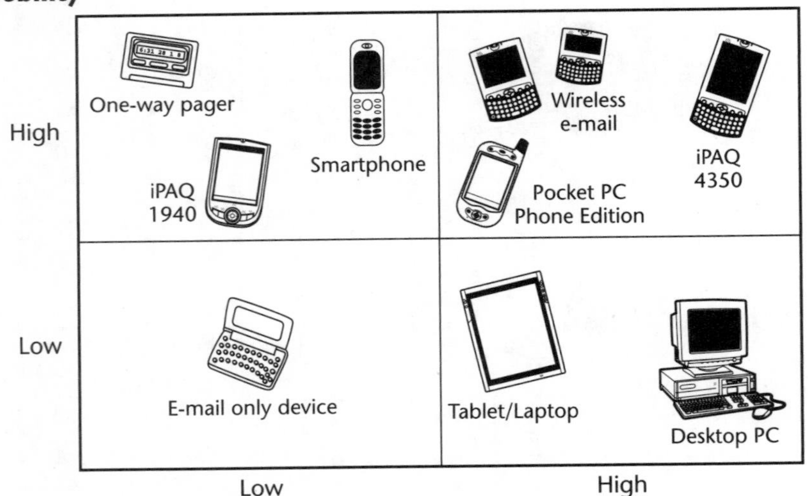

Figure 1.4 Mobile or wireless devices?

In Figure 1.4, CGE&Y performed a study that indicates various devices and where they might fit in a spectrum between requiring a wireless connection to simply acting as a stand-alone mobile device. The most obvious nonwireless, nonmobile device is the desktop PC, because a wired connection is typically required and carrying your PC with you is an unlikely scenario. On the other end of the spectrum is a wireless e-mail device or a Pocket PC, both of which are easy to take along with you wherever you go and usually require some wireless capability for their applications to function. It is important to note, however, that a Pocket PC type of device also fits well as a fully functional nonwireless mobile device, whereas a nonwireless Palm III is primarily a nonwireless mobile device.

Always On

The term "always on" refers to the characteristic of a wireless device to have to have the capability to constantly send or receive data. In other words, it's not necessary to initiate a dial-up connection with a modem, but the connection is constant. "Always-on" technology is quite often referred to in the third generation (3G) world of wireless systems. In case you're wondering what we mean by 3G, take a look at Figure 1.5 for a brief overview of the timeline of the generations of wireless technologies.

Without going into great detail on each of the generations of wireless technology, it's fairly easy to see in Figure 1.5 that each generation enables new functions and experiences for the mobile user. One of the best examples of "always-on" technology is an E911 (Enhanced 911) service. The basic idea behind E911 is to locate an individual who requires emergency attention. The device that is carried by the person needing attention transmits a signal that enables the home office to find the person within 50 and 300 meters. Not bad, but the device must always be on for the patient to be found.

Remote Access

An attribute of mobility that probably comes to mind first is the ability to access information while away from the office or home where the comforts of the wired desktop PC reside. Most devices, even a laptop, are small enough that they can be carried virtually anywhere. Even better, if my device is capable of a wireless connection, I can now access my data almost anywhere.

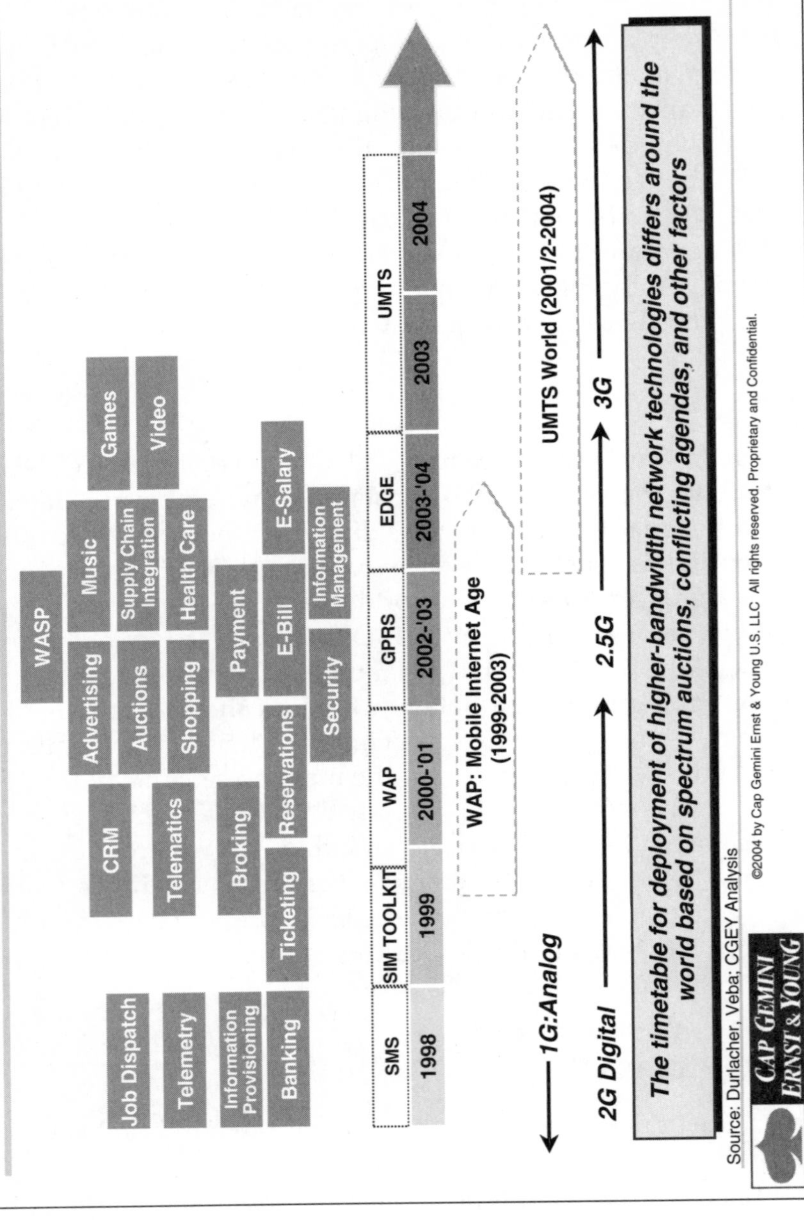

Figure 1.5 The generations of wireless technologies.

Localization

Localization refers to the ability to locate your customers and provide them services based on their current location. Sound familiar? It should, because localization was mentioned in the previous section under location-based services. Localization is another attribute of wireless devices that provides additional services to the wireless user. An example of such a service goes back to sales force automation application. Again, the sales force is typically a mobile workforce traveling from city to city routinely. Wouldn't it be nice to know if another member of your sales team is in your area so that he or she may be able to assist on a presentation? There has been a great deal of talk about privacy and localization. Do I want just anyone to know where I am at all times? It really depends on the customer and what the service is. Some consumers don't mind receiving coupons for coffee while passing the local Starbucks. E911 services are dependent upon localization to work properly. The bottom line is that localization can be turned off. It's certainly not a requirement. The potential, however, is very high, especially for the enterprise, because it can target customers based on their likes, dislikes, and location.

Ubiquity

Whether you carry a wireless or a nonwireless mobile device, you always benefit from the ubiquitous personality of mobility. Any mobile device has the ability to perform work anywhere and anytime. Whether you're walking in the park and need to check your work schedule or you're receiving an e-mail message on your mobile phone, mobile devices go anywhere you go.

Personalization

As people, we are all unique beings who like and dislike different things. I like to listen to blues and rock music, whereas the next person may like jazz. I like to eat Italian food, whereas the next person may like Greek. It's an ever-changing and evolving world we live in that constantly gives us new things to enjoy and new places to go. The world of the Internet also strives to provide an adaptable experience that makes us want to come back.

The idea of personalizing the Internet experience opens the door to commercialization. If I know what my users are interested in buying, I can market specifically those products to them, instead of mass marketing a million different products to them. A good example of personalization is the very popular Amazon.com site. Once you log into the site and tell

Amazon a few things about yourself, the shopping experience begins. As I shop throughout the various electronic malls at Amazon and make various purchases, my personal likes are cataloged for future reference. For instance, I bought a classical music CD and a comedy DVD movie. In my future shopping with Amazon.com, they will provide me with links or match the new comedy movies to my user profile. Additionally, I won't be bombarded with online commercials for every ad under the sun.

Now, let's take personalization to the next level and provide a mobile spin to it. One of the advantages of location-based services is the ability to pick and choose what kind of alerts I would like to receive. For instance, when I'm in Chicago, I'd like to be notified of a good blues bar when I'm driving near an area rated high on Zaggat's list of blues bars. Because I'm also a fan of Italian food, I'd also like to be notified when I'm near a highly rated Italian restaurant. You can imagine the possibilities if we can learn more about our customers. The customers are happy because they feel like they are being treated as individuals and not just part of the mass market. The business is happy because customer satisfaction and the percentage of return business are high.

Proactive Push

One of the characteristics of mobility (notice that we said mobility and not just wireless technology) is that we now are holding a device that can accept information anytime and anywhere. Thus, our business partners now have the ability to send us e-mail, instant messages, alerts, and other data. This is called pushing the messages or data to the device as opposed to the device requesting the information. Proactive pushing works quite well in location-based services applications because it's a messaging-based service. Pushing messages to our customers is one of the earliest innovations in mobility. Small message services (SMS) have been around for quite some time, and in fact, is the most popular mobile service in Europe. SMS messages are basically like mini-e-mail messages that can be sent to and from any SMS enabled device, such as a mobile phone. The simplest example of pushing data to a device is an instant message sent at the same time every day with the day's weather. There are numerous services, such as MSN Mobile and Yahoo mobile, that provide instant messaging to capable mobile devices. Proactively pushing messages works quite nicely in the business world as well. Supply chain optimization applications that monitor product shipments and notify the proper management personnel of breaks in the supply chain through hot alerts, are an example of push technology. We'll cover proactive push applications throughout this book.

Summary

As we bring this first chapter to a close, let's quickly review some of the key concepts. You learned that the terms wireless and mobile can actually have very different meanings when applied to the world of mobility. Generally, when we talk about wireless devices or wireless technology, some sort of direct connection to a network is required for the application to function as intended. The wireless connection can be through a company network, usually a firewalled private network, or through a public Internet connection through a wireless service provider. An example of a wireless application is Web browsing through Internet Explorer or some browsing device. The term mobile spans the realm of wireless or nonwireless applications because in many instances it's not necessary to be fully connected to a network to perform work. An example of a mobile application is an intercompany address book. The names in the address book are available anytime and anywhere, whether you are connected to a network or not. Once the user syncs or initiates a wireless connection to a network, the address book is updated with the most current changes. Finally, we covered the various attributes of mobility and what makes these characteristics unique to mobile and wireless applications. The next chapter covers the growth factors of mobility.

CHAPTER

2

The Driving Forces of Mobile's Growth

Adam Kornak

Introduction

Now that we've defined the similarities and differences between mobility and wireless, it's important to understand how to apply them to a real business application. Clients today generally come in two categories: consumers (retail customers like you and me and the business customer (corporations, sole proprietor, and so on). In the world of e-business, we're accustomed to thinking about the customer through a relationship-based approach. The most common relationships are the following:

- **Consumer-to-consumer (C2C).** The Internet has been wildly successful at providing a completely new marketplace for consumers to market goods and services to fellow consumers. eBay is a perfect example of customers selling to customers through an online auction-based arena. I can buy or sell anything, from a Monet painting valued at 10 million dollars to an American flag key chain selling for a mere $1.50. There are no real suppliers, middlemen, or producers in this scenario. It can be compared to a many-to-many supply relationship. The buyer and seller don't need to have a special relationship with each other to buy the offered product. As a seller, I can market my product in any way that's easiest for me, as long as I respect the rules of the online host, which in this case is eBay. We'll talk more about how mobility fits into the C2C world a little later in this chapter.

- **Business-to-consumer (B2C).** Another very common e-business scenario is the business-to-consumer (B2C) relationship. In our daily lives, it's almost impossible to imagine a moment when we are not faced with some sort of purchasing decision. All of us are consumers of goods that are produced by a retailer who attains those goods through some supplier relationship. That relationship is known as a business-to-consumer relationship. A typical example of B2C is buying your favorite music CD or DVD movie on the Web through an online retailer such as Amazon.com. Anyone who purchases the items on Amazon.com is the consumer, and Amazon.com is the business or retailer selling the product to the consumers.
- **Business-to-business (B2B).** Still another common method of leveraging the Internet for commerce is the business-to-business (B2B) relationship. Regardless of the type of product a company sells, they must attain their goods to be sold from a supplier or make it themselves. In the B2B example, the buyer (retailer) usually buys the product from a supplier (wholesaler). If we go back to the Amazon.com example, we know that Amazon is simply the middleman and buys their products to be sold from wholesalers with whom they've established relationships. The supplier of Amazon's goods manages the supply side of the B2B equation. Ultimately, the goods end up in the consumer's hands, through, for example, the purchase of the DVD or CD.
- **Business-to-employee (B2E).** Organizations have been quick to realize the ubiquitous quality of the Internet and intranets by creating new opportunities for their employees to access intercompany data. Human Resources departments are using company intranets more and more to pass along benefit information, online forms, 401k information, and terabytes of other information that is making an employee's job easier.
- **Government-to-citizen (G2C).** Finally, federal, state, and international governments may not have been the first adopters of the Internet for commerce, but let's not forget that the Internet was developed for government and educational use in the first place! Government agencies such as the FBI, CIA, IRS, and many, many others are leveraging the power of the Internet to pass important

information to us, the citizens. Everything from IRS tax forms to information about the most wanted criminals is available online for all to use. The purpose of these systems is primarily for information and less for actual barter exchange and commerce.

As you'll see, many of the early generation mobile applications are really just an extension of our wired desktops, so when we discuss the various commerce relationships such as B2B or B2C, those relationships still exist in the unwired world.

This chapter goes into greater detail on why mobility is so important and what is driving its growth. We'll cover the consumer drivers as well as the business drivers of mobility. Finally, we'll review some of the trends in the wireless industry.

Consumer Drivers

Imagine yourself driving to work on a normal workday minding your own business and doing your best to get to work on time. You decide that you'd like to catch the latest stock market news along with an update on your financial portfolio. While driving, you speak into your voice-activated Telematics monitor and ask for the "latest financial market news." The system responds back with "Dow Jones Industrial Average is up 52 points, Nasdaq is up 23 points, and S&P is up 3 points. Total portfolio value up 3 percent, would you like to buy or sell holdings?" You respond with a "no thank you" and continue driving. As you pass the local Starbucks on the way to work, your monitor beeps and sends you an instant message that Starbucks has a sale of $6.50 on a pound of Guatemala coffee. You wonder how Starbucks knows that Guatemalan is your favorite brand of coffee. You stop and pick up a pound for yourself along with a latte. As you stroll into work, your boss sends you a hot alert on your mobile phone to remind you of the board meeting that you are to attend that morning.

The previous scenario touches upon many of the services that consumers, like you and me, are already starting to experience. Telematics—location-based services, personalization, and voice-command-activated mobile services—is a very compelling way to drive the consumer to new avenues of experiencing commerce and convenience in the mobile world. This section of the book covers how those mobile services and many others are changing the way consumers are shopping, working, learning, and living.

The Quest for Convenient Communication

An interesting statistic from a recent Forrester research study predicts that by the year 2004 B2C sales in the United States will reach a staggering 3.2 trillion dollars! (see Table 2.1).

There's no doubt that online shopping has hit big and is only expected to grow dramatically over the next few years. Those same services are extending into the realm of mobility, as consumers are demanding and obtaining quicker and easier ways of accessing the Internet. As Table 2.2 indicates, the demand for the wireless Internet is also growing exponentially.

Table 2.1 Predictions for Global Regional B2C Sales in 2004

WORLD REGION	TOTAL (DOLLARS)
United States	3.2 trillion
Asia-Pacific	1.6 trillion
Western Europe	1.5 trillion
Eastern Europe, MEA	68.8 million
Latin America	82 billion

Source: Forrester Research

Table 2.2 M-Commerce Connections (000s), 2003–2008

	2003	2004	2005	2006	2007	2008
World	51,334	91,712	145,408	210,548	282,434	350,849
North America	2,095	5,130	12,349	26,521	44,616	58,956
Latin America	2,368	5,283	9,943	16,225	23,898	32,536
Western Europe	26,594	43,428	60,286	76,699	92,300	106,605
Eastern Europe	2,223	4,993	9,032	13,893	18,994	23,863
China/India	1,659	4,600	10,391	16,521	23,456	30,971
Asia-Pacific	14,818	24,861	36,983	50,701	65,151	79,631
Middle East and Africa	1,578	3,417	6,424	9,990	14,019	18,287

Source: Ovum

Table 2.3 M-Commerce Revenues ($million)

	2002	2003	2004	2005	2006	2007
World	51	114	203	317	450	582
North America	2	7	19	45	87	131
Latin America	1	2	4	7	11	16
Western Europe	30	65	108	154	200	242
Eastern Europe	1	2	4	7	10	13
China/India	1	2	5	9	14	19
Asia-Pacific	16	35	60	89	120	151
Middle East and Africa	1	2	4	6	9	12

Source: Ovum

Without a doubt, there are many factors that play into the demand for mobility. Consumers today are presented with countless ways of leveraging technology in living their daily lives. Mobility has offered another avenue for consumers not just to purchase goods and services online, but also to conveniently make those purchases on the go, anywhere, at anytime. As Table 2.3 indicates, consumers throughout the world are adopting this method of commerce. Let's not stop there; mobility is offering services that could never exist in a wired world, such as E911 and mobile wallets. Soon, it might not be necessary to carry paper money, because electronic money can be made available on smart cards or through electronic wallets that go with you everywhere. Regardless of the popularity of these mobile services, certain challenges seem come up time and time again that drive a consumers decision to go mobile (see Table 2.4).

Table 2.4 Obstacles Preventing Consumers from Adopting Mobile Commerce

OBSTACLE	PHONES	PDAS
Credit card security concerns	52%	47%
Fear of "clunky" user experience	35%	31%
Don't understand how it would work	16%	16%
Other	11%	13%
Never heard of it before	10%	12%

Source: Forrester Research

Those challenges include:

- **Security.** We all remember the first time we made an online purchase using our credit card. After you bought your product online you were probably immediately asking yourself questions such as "Did my credit card number make it through?" or "Were there any prying eyes that grabbed my credit card information?" or better yet, "Will I ever get the item that I ordered?" Having the comfort of knowing that personal financial information is safe from hackers is one of the biggest concerns today. According to the Nation Consumers League, the amount of money consumers lost in consumer fraud in 2002 was over 14 million dollars, with more than 13 million from online auctions alone. "The average loss per person rose from $411 in 2001 to $484 in 2002." With those kinds of statistics, it's easy to see why security is high on the list of concerns that consumers are facing when on buying or browsing on the Net. Suffice it to say that there is a great deal of measures being taken by organizations to make the wireless online buying experience a safe and secure method of shopping. Some of those methods include cryptography through confidentiality applications, server and client authentication, digital signatures, public key encryption, and digital certificates. For more information on security in a mobile environment, see the book *The CGE&Y Guide to Wireless Enterprise Application Architecture* by Adam Kornak and John Distefano (John Wiley & Sons, Nov. 2001).
- **The Customer Experience.** We, as consumers, have very high expectations for new technologies. Mobility requires us to frame our experience in a completely different way than we've been accustomed to. The average screens size for most mobile devices is merely an inch or two in width. That leaves little room for error for content designers. If our experience on the wireless Web does not meet our expectations, it's just as easy for us to switch back to our wired desktops. It's easy for Web designers to make the same mistakes that early Internet providers made by focusing on the technology and not the customer. In the days when the Web was in its infancy, anyone and everyone was in a fury to launch their company's Web site, giving little thought to the strategy or content of the site. That same idea is being applied to many wireless Internet sites that were launched simply to establish a presence. Now that consumers have the option of over a hundred different mobile devices and virtually hundreds more sites to browse, it's even more critical to focus on your customer's experience.

- **Devices, Devices, Devices.** With the last count of over 200+ mobile devices, the choices consumers need to make in purchasing a mobile device are frightening. The mobile device environment is more dynamic than ever, making the decision-making process for consumers as exciting as ever, yet potentially confusing. It's not uncommon to purchase a mobile phone or wireless PDA, and see that same device upgraded one month later. There doesn't appear to be an end in sight for mobile device innovation. With that in mind, device providers need to keep their consumer in mind more than ever. It's almost impossible to be everything to everyone, but many providers have almost succeeded in designing the perfect mobile device. We'll discuss the various mobile devices later in this book.
- **Lost or Stolen Devices.** The beauty of purchasing that shiny new desktop PC was that usually it was big enough that theft was of little concern to the home user. And, you'd have to be a pretty sorry customer if you actually lost your new Pentium PC. Mobility has changed everything about those formulas however. Mobile phones and PDAs are now smaller than many men's wallets and more likely to be lost or stolen. Because mobile phones are used as a means of communication, not just commerce, we find ourselves accessing those devices constantly in our daily lives. Replacing a mobile phone may not be as traumatic as we think, because service providers can quickly replace them with a mere phone call. But, the loss of convenience or data stored on the device certainly weighs into our decisions in how we use mobility today and in the future.

Right Here, Right Now—Consumer Expectations

Customer lifestyles, careers, families, and expectations as a whole are drastically changing the way enterprise players are viewing the marketplace for goods and services. The "speed of change" is not just asking, but demanding, that producers of goods and services react to consumers' demand for anywhere, anytime services. Our customers are working longer hours, traveling more (for business and pleasure), needing to meet tighter deadlines, and basically not taking "no" for an answer. Mobility is certainly not the answer to all those customer needs and wants. But, it is offering solutions to problems and questions that could not be answered some years ago.

The Device as a Personality Extension

As human beings, we naturally want to be liked by our friends, coworkers, family, and in general anyone we meet. It's an inherent need of humankind to appear attractive and have a personality that people enjoy being around. Mobile device manufacturers understand that need and recently have applied those attributes to the design of their products. Since the dawn of the first mobile phone, or first generation analog phones (see Chapter 4), the only purpose of the mobile device was simply to make a phone call. Frankly, that's about all the analog phone could do at the time! Devices were large, bulky, and not terribly easy to lug around, not to mention rather unattractive. Generally, users did not carry their phones on their person for 15 hours a day or more as we do today. It simply wasn't practical to do so. In addition, most people looked at you rather strangely if you carried a huge phone on your belt all day long or during meetings.

All that has changed, however, since mobile devices have shrunk to a size smaller than a wallet. Mobile devices have also grown in usage to an alarming extent. A recent study indicated that the number of mobile phones in Sweden outnumbered the number of wired phones! The fact that the usage of mobile technology has grown at such an enormous rate and that users now carry their mobile devices almost 24 hours a day presents new opportunities to design devices as an extension of our personalities. A peek at any of the popular device manufacturer Web sites will give you a small idea at the diverse selection of colors, sizes, and styles of mobile phones and mobile devices available. Whether you like red phones, blue phones, flip phones, or color phones, not to mention PDAs and any number of wireless devices, a new market has been opened up through a new art form through mobile device design.

Another area that has shown an increasing amount of popularity is mobile phone ring tones. Mobile phones today generally come equipped with a standard set of ring tones that users can choose from. Device manufacturers realized quickly with the growth of phone usage that distinguishing between ring tones was becoming increasingly more difficult. How many times can you recall strolling through a public place listening to any number of cell phones jingle to the same tone and seeing everyone in your vicinity checking their phones at the same time? Service providers and device manufacturers realized an additional need not only for the personalization of ring tones, but also for a distinction among other common rings. New Web sites are developed almost daily that allow customers to purchase and download ring tones for their phones, ranging from popular rock/pop songs to hit movie themes.

As you can see, as mobile devices become more and more a part of daily existence, they must also be pleasant and stylistically appealing for customers to want them. Just as with any product that we've grown accustomed to using on a daily or hourly basis, we want the right to choose the exact model and style because it not only represents our likes and dislikes but also is truly an extension of our personality.

Fun and Games

A very exciting and quite rapidly growing area within the wireless space is entertainment on the go. The wireless gaming experience is obviously a bit different from what many video game fanatics are accustomed to. Mobile devices are equipped with screens ranging in size from a postage stamp (cell phones) to that of a larger Post-it note (PDAs), which can somewhat limit the capacity for a rich customer experience. Traditionally, the qualities that make a video game fun and exciting are the graphics, sound, and speed of the game, all of which are qualities that most of today's mobile devices can't compete with. However, mobile devices and wireless networks are rapidly evolving and those limitations are melting away. An interesting statistic gathered by Ovum (a leading research group) indicates that total revenues from wireless entertainment will increase from only $125 million in 2001 to over 4.4 billion in 2006. Not too shabby. Depending upon where you spend most of your time and which wireless network you primarily use, the quality of your gaming experience will be drastically different. For instance, a typical wireless game in the United States might have a connection speed of 14.4 Kbs (kilobytes per second) with the potential to be in thousands of colors. Typically, with a mobile phone connection, most of the games take a fair amount of time to download to your device and require a constant connection to be playable. Some examples of these games include Solitaire, Craps, Blackjack, and various adventure games (see Figure 2.1).

Figure 2.1 Wireless games on a mobile phone.

On the other hand, the Palm and Pocket PC operating systems have succeeded in producing a very different gaming experience. Figure 2.2 illustrates a screen shot from a popular golf game on the Pocket PC operating system from Microsoft.

As you can see, the graphics are pretty impressive for a handheld device. iGolf is an example of a peer-to-peer (P2P) networked game, meaning that you can connect directly to another Pocket PC device and play someone online. You can also play while in disconnected mode, so a wireless connection is not required.

Figure 2.2 iGolf on a Pocket PC.

A New Channel for Existing Entertainment

Another exciting category in wireless gaming utilizes a technology discussed earlier in the book, called *location-based services*. A company in Sweden called It's Alive! has developed a game called BotFighters. The concept of the game is to recruit an electronic robot that acts as your character and virtual fighter for the game. Using location-based tracking with a mobile phone, participants track down other robot fighters and shoot them when they are in range. Of course, the shots are virtual as well. Points are tallied for shooting opponents. Players have been known to travel over 30 miles or more to find and shoot their opponent with their virtual weapon. Now that's dedication! These and many other channels of opportunities are being developed as technologies like location-based services are taking off.

Another very profitable channel for wireless entertainment is online gambling. According to *World Gaming*, 73 percent of online gaming is spent in Internet casinos and sports betting sites. A recent report from Datamonitor anticipates that total revenue from mobile gaming markets in the United States, Europe, and Asia-Pacific will grow from about U.S. $950 million in 2001 to $17.5 billion in 2006. With those kinds of statistics, the opportunity for gaming and entertainment companies is vast.

The revenue opportunities for wireless gaming providers can come in many forms. Some of the key drivers are:

- **Wireless Internet and monthly membership usage fees.** Basic monthly usage fees for mobile voice or wireless Internet usage. Some plans charge by the hour, whereas others are bundled with voice services.
- **SMS (small message service) text message cost.** A fee based on the number of SMS messages sent and received through the mobile device.
- **Shrink-wrapped products.** Revenue earned by games that are purchased through typical retail channels, such as online shops or computer stores. The game requires installation onto a device, such as a Palm or Pocket PC.
- **Gambling revenues (where permitted).** Monies generated through wireless online gaming sites.

WHAT IS 3G?

3G, or third generation, is an ITU (International Telecommunications Union) specification for the next generation of mobile communications technology. The first generation of mobile technology was analog communication, and 2G was digital PCS. 3G will improve bandwidth significantly—up to 384 Kbps when a device is stationary or moving at pedestrian speed, 128 Kbps in a car, and 2 Mbps in fixed applications. 3G will work over wireless air interfaces such as GSM (Global Standard for Mobile Communications), TDMA (Time Division Multiple Access), and CDMA (Code Division Multiple Access). The new EDGE air interface has been developed specifically to meet the bandwidth needs of 3G. See Chapter 4 for more detailed information on 3G and other wireless technologies.

These and many other new revenue drivers are quickly changing the way the video game and entertainment industry is viewing wireless technology. A unique quality of video game entertainment seems to be that it is only slightly affected by the downturns in the economy. When the economy is flat, consumers turn to entertainment for a low-cost mechanism of relaxation. On the other hand, in an upturn of the economy, consumers spend even more on avenues of entertainment that they always enjoyed. The next section discusses some new types of entertainment in the works.

New Types of Entertainment

New opportunities are quickly arising in the wireless gaming and entertainment space. Certainly a factor in the development of these games is the migration of wireless networks. It's quickly becoming apparent that 3G (third-generation) wireless networks are advancing upon us quickly.

By the end of 2004, most large telecommunications carriers will have the bulk of 3G networks implemented. One of the advances in the wireless gaming experience that is brought on by 3G networks is streaming video and music. With increased bandwidth, entertainment seekers will have the opportunity to play online video games with the same speed and crispness as a standard desktop broadband connection. Additionally, as mobile devices are transformed to accommodate the new bandwidth, a whole new market for wireless gaming will be established.

Business Drivers

Now that we've looked at the consumer drivers of mobility, let's take a look at the other end of the spectrum—the business drivers of mobile applications. While mobility offers very exciting opportunities for companies, it is not necessarily an application that everyone should rush out and implement immediately, at least not without a strategy. Adequate planning should be a very high priority prior to implementation. Additionally, companies should take advantage of the unique opportunities mobility provides and not just rush to porting an existing Web presence to mobile devices. Mobility is about access to information regardless of time and location—access provided to employees within the enterprise, or to constituents such as customers, suppliers, or business partners outside of the organization. As a start, the mobile strategy should address issues that follow well-defined business objectives by following a structured strategic planning approach that consists of the following stages:

1. Developing an overarching wireless vision
2. Defining the organization's wireless direction/strategy
3. Creating a prioritized portfolio of wireless initiatives
4. Charting a roadmap for the deployment of high-priority initiatives
5. Validating the approach via a proof of concept
6. Implementing the strategy
7. Monitoring and adjusting the strategy

We will discuss the strategic planning process in detail in Chapter 6. Just to give you a hint of some of the considerations that companies must make, we've listed some important questions that companies should ask themselves prior to a mobile system implementation. These include:

- What are your business objectives?
- Who are your customers and how will they benefit from mobility?
- Who are your competitors and what are they currently doing in the mobility or wireless space?
- What internal skills do you currently have and where do you need help to achieve your goals?
- Have you identified organizations that you would potentially partner with in your mobile commerce endeavor?

- How will you measure the success of your mobile application implementation?
- Do you have a marketing strategy and communication plan?
- Do you have a release plan?
- How will you support your mCommerce site?

Figure 2.3 illustrates additional questions that companies should address around mobile commerce offerings.

Many of these questions are not unique to mobility. You'll find that you may have asked yourself these types of questions with any significant system implementation. It is, however, very easy to forget the basics and make the rush to the bleeding edge. The drawbacks to a rushed approach to implementation are failed implementations, a lack of business direction, a lack of usability, customer confusion, and ultimately failure—dropping mobility altogether. This section discusses the various business drivers of mobility in any organization as well as uncovers real applications that can help you get started with mobility in your place of business.

Cost Efficiencies

One of the questions you should ask yourself is, "How will you measure success of your mobile application implementation"? This section takes a stab at answering that question through examples of the potential cost savings that you might expect to achieve through various mobile applications. It also reviews sample cost-benefit analyses that were performed as part of specific CGE&Y mobility case studies.

COST-BENEFIT ANALYSIS AND RETURN ON INVESTMENT STUDIES

It's important to mention that an organization will attain a specific measurable return by implementing mobile technologies or using our business case analyses. There are many factors that go into developing a cost-benefit analysis, and the elements of each study can be very different for each organization. This book simply outlines a general rule of thumb that CGE&Y has determined through rigorous cost-benefit studies.

More Broadly, Companies Should Address Several Critical Questions Surrounding Mobile Data Offerings

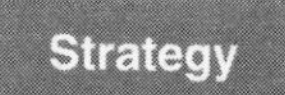

- Who are the target customers for mobile services, and how will loyalty, community, and switching costs be established?
- How will companies manage the mobile customer experience and capture customer data across multiple channels and interactions?

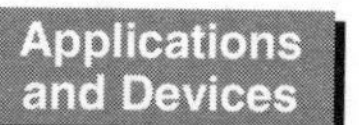

- What applications and content are required by target customers?
- How will companies foster and manage the network of alliances and partnerships to deliver compelling mobile data solutions to target customers (e.g., app. developers, device mfgs., content providers)?

- What business processes should be modified to support the mobile offering (e.g., network management, customer service, billing)?
- How will companies integrate applications and manage service quality across carriers, enterprises, content providers, and devices?

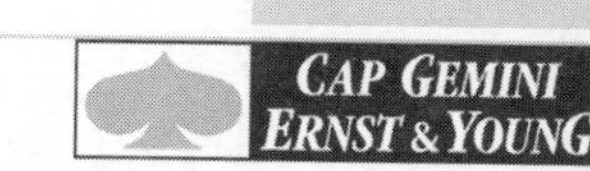

Figure 2.3 Critical questions companies should ask.

Wirelessly Enabled Employees

One of the first, and the safest, ways that organizations can begin using wireless technology is by enabling their workforce. It's fairly safe, because it's much easier to pilot a mobility program with your own employees and attain qualified feedback about where you can improve as opposed to losing out with an expensive implementation with a client. There are many mobile applications discussed in this section, including some in the B2E (business-to-employee) space that we discussed in Chapter 1. The approach for this section is to look at the benefits of a mobile workforce and then drill down into the mobile applications that are currently considered cost-efficient enablers. As such, the key benefits of a mobile and a potentially wireless workforce are:

- Reduced traveling
- Streamlined data entry
- Scheduling effectiveness
- Administrative management

The following sections cover each of these areas separately, while also providing a specific mobility application representative of each.

Reduced Traveling

A great deal of time and money go into traveling to meet with clients to make presentations, perform work, attain knowledge, and monitor progress; traveling is simply a requirement for many jobs. Certainly not every role within an organization requires travel; however, more and more employees are becoming a mobile workforce. In other words, it's becoming a necessity to travel to perform our jobs. You may recall a past job interview with either your current company or a past company that listed as a job requirement, travel required. If you've ever worked as a sales team member, quite often that is the case. If so, you're considered a perfect candidate for workforce mobility. Some of the typical roles that we've seen within an organization are:

- **The sales force.** There's no doubt that the sales team comes to mind first when thinking about a mobile workforce. The sales team undoubtedly is the first and most seen face to the customer. CRM (customer relationship management) mobile tools are an example of mobile applications.

- **Field service or customer service technicians.** Most organizations today supply some sort of customer services for their clients. There are many levels of customer support, varying from a single-person help desk to a full-fledged customer management call center, all of which can benefit from mobile applications.
- **Executive and upper-level management.** CXO-level employees, vice president's, directors, and so on, all make business decisions based on accurate information and data. Sometimes that means reviewing company financials, whereas other times it's forecast data required for board meetings. Making that data available in real-time through mobile office applications can be a tremendous advantage in an executive's decision-making process.
- **General Office Workforce**. Believe it or not, anyone who works in the confines of a conference room or cubicle is also a candidate for mobile applications. Anyone who spends time servicing other employees and staff within a brick and mortar foundation can realize value from wireless local area networks.

SFA (Sales Force Automation)

Accurate, real-time sales pipeline information is the most important asset a salesperson, manager, and executive can possess. This critical information enables the entire sales organization to make fact-based decisions that will maximize sales results, and in addition, establish performance expectations for their most important stakeholders—customers, shareholders, and the financial community. sales force automation (SFA) technology is the most effective and efficient means to capture this valuable information.

Traditional SFA tools typically require a desktop client, in the case of a client/server-based architecture, or a Web-enabled browser client for an Internet-based approach. In either case, a desktop is required with a live connection to a client database residing somewhere in the background. Some of the more common functions within a SFA application can include:

- Forecasting or business planning
- Knowledge management
- Account management
- Territory management
- Opportunity management
- Proposal definition

These are just a few of the growing list of functions that sales teams can access in an SFA tool case. So what are the benefits of mobilizing your SFA tools? As we mentioned previously, the sales force is a natural fit to leverage the powerful tools that mobility provides. Some of the key benefits in a mobile SFA application include:

- Avoids unnecessary reentry of information from paper-based forms; ensures accuracy
- Provides faster turnaround time on quotes
- Provides comprehensive price and product comparisons to assist in selling process
- Reduces/eliminates unnecessary paperwork
- Increases sales agent efficiency—reduces cycle time and increases sales call potential
- Increases sales agent effectiveness—real-time access to information
- Increases sales due to more effective cross-selling
- Allows sales team to always be "connected" and have access to critical information
- Increases sales per agent
- Allows client meetings to be shorter and more informative
- Provides targeted information
- Improves the customer experience

A typical example of a mobile SFA application business process is seen in Figure 2.4.

As you can see in Figure 2.4, the salesperson can now submit proposals while in front of the customer, saving value-added time. The summary of a CGE&Y cost/benefit analysis shown in Figure 2.5 provides an example of how mobility can reduce costs and potentially increase revenue with sales force mobility tools.

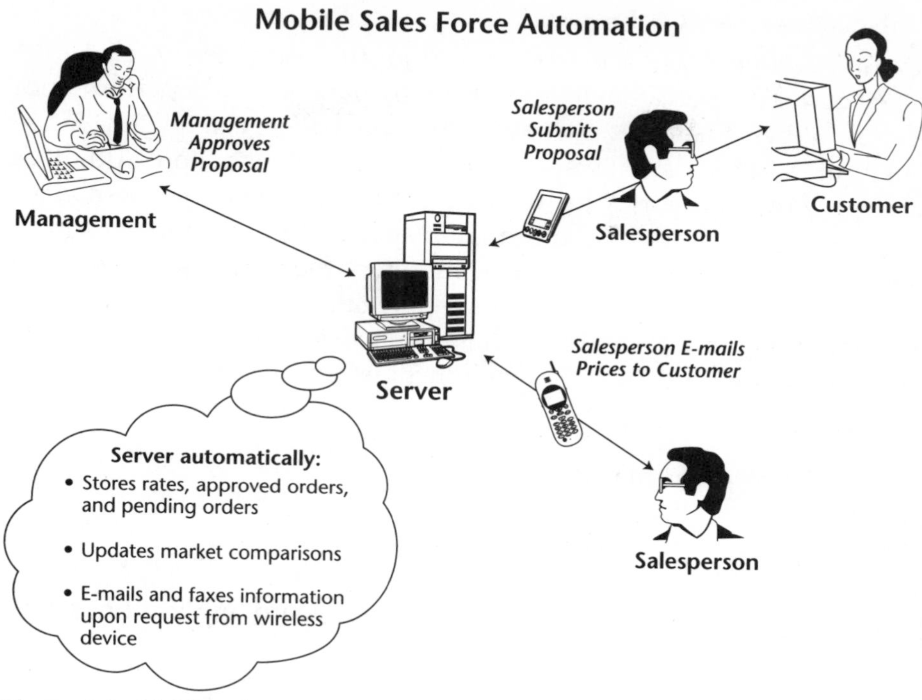

Figure 2.4 Mobile sales force automation process.

VALUE PROPOSITION:		Less Time Spent on Paperwork		
MOBILITY FUNCTIONS:		1. Process invoices, billing, and capture signatures 2. Mobile office functions: e.g., expenses, timesheet		
COST SAVINGS:				Example
	#1	# of orders per day	2	5,000
		Average cost/hour per field worker		$40
		Time saved by inputting info onsite (e.g., service requests, orders, contacts) (minutes)	2	10
		Subtotal #1		$7,666,667
	#2	# of field workers		1,000
		Time saved in administrative tasks/day (e.g., timesheet, expenses, client info) (minutes)		10
		Subtotal #2		$1,533,333
		Total Cost Savings per Year		**$9,200,000**
REVENUE GENERATION:		Extra service calls/week		1
		Average revenue per service call		$200
		Total Extra Revenue per Year		**$10,000,000**

Figure 2.5 Value proposition: less time spent on paperwork.

Executive Management Dashboard

An *executive dashboard* is a management tool that provides an executive with key corporate performance measures. By having critical performance indicators at their fingertips, executive decision makers can quickly gauge the state of a particular business process and initiate action. Although capable of showing metrics for the company as a whole, the dashboard usually focuses on providing information specific to the company's individual business units. The dashboard monitors the performance of such business units, and displays the data in aggregate form, sparing the dashboard user from getting lost in the data. An executive dashboard mobile application can provide the following functions:

- Financial performance
- Operational performance
- Employee productivity metrics
- Customer satisfaction metrics

A more detailed case study on an executive dashboard application is covered in Chapter 10.

Field Service Tool

Field service personnel and technicians can benefit from mobility in similar ways to the sales force teams. Field technicians are also required to spend a great deal of time away from their offices. Quite often, a field service technician's office is his or her car. Like the sales force, the field force team must service the customer's needs. Whether that service involves taking new customer orders, tracking of existing orders, or monitoring systems in the field such as meter readings. Figure 2.6 illustrates clearly the benefits of FSM (field service mobility) applications.

From a return on investment perspective, FSM applications have been shown to provide measurable benefits over time, as Figure 2.7 illustrates.

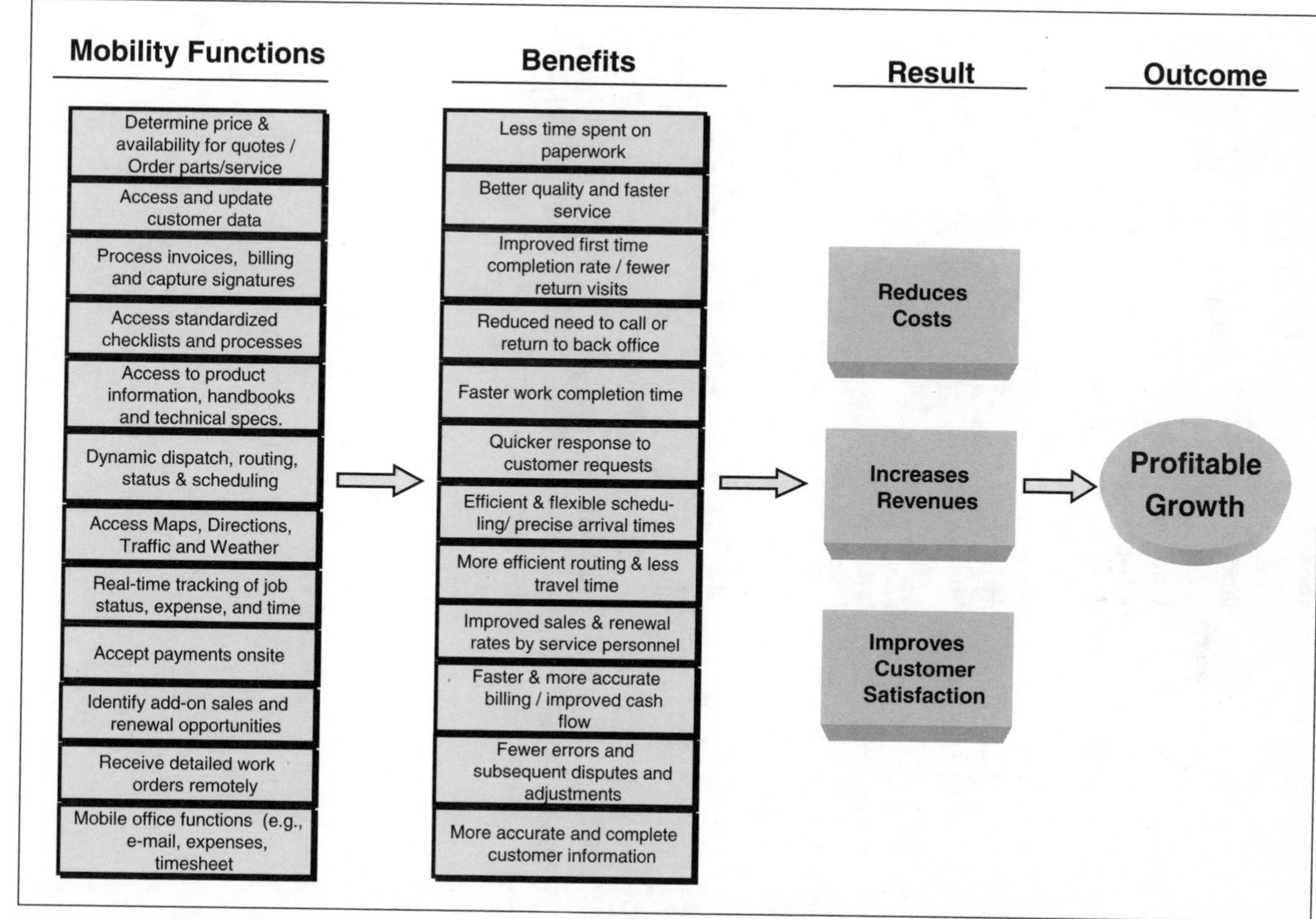

Figure 2.6 FSM functionality reduces costs, increases revenues, and improves customer satisfaction.

		Client Calculation		Example
VALUE PROPOSITION:	Better Quality and Faster Service			
MOBILITY FUNCTIONS:	1. Access historical customer data 2. Accept standardized checklists and processes 3. Access to handbooks and technical specs			
COST SAVINGS: #1	# of repair orders per day	0	2	5,000
	Average cost/hour per field worker			$40
	Time saved /repair order by having accurate customer information (minutes)...	0	2	10
	Subtotal #1			$8,333,333
#2/3	Time saved by having better technical information (minutes)			20
	Subtotal #2/3			$16,666,667
	Total Cost Savings per Year			**$25,000,000**
REVENUE GENERATION:	# of field service workers			15,000
	Extra service calls/week	0		1
	Average revenue per service call	$0		$200
	Total Extra Revenue per Year			**$13,800,000**
CUSTOMER SATISFACTION:	Fewer errors and subsequent disputes and adjustments More consistent service			

Figure 2.7 Better quality and faster service through FSM.

A more detailed CGE&Y FSM case study is covered in Chapter 10. As you'll see in the next section, the office infrastructure also provides many optimization opportunities for mobility.

Wireless Local Area Networks

Since the inception of the 802.11b and the 802.11a IEEE standards, wireless LANs (local area networks) have become an efficient means of accessing internal company information, especially with wireless conferencing centers and wireless workspaces. All too often, employees' days are spent in conference rooms attending meetings. Wireless networks provide an efficient means for mobile device users to access any company applications that would normally be provided over a company LAN. With wireless networks providing data speeds in excess of 5 GHz, it's easy to see why wireless networks are becoming popular.

Revenue Growth

Mobility generates revenue growth opportunities that never would have been imaginable before. Leveraging mobile technologies successfully is all about identifying increased revenues or decreased costs, as with all other investments. The profits are normally achieved by increased internal efficiency and improved or new customer services. It is fair to say that companies not using those possibilities will not be competitive in the future. Clients, business partners and employees will soon expect accurate real-time information from all types of companies. This section describes some of the revenue opportunities companies can take advantage of through mobile commerce applications.

Wireless as a Revenue Driver: mCommerce

Mobile commerce (m-Commerce) will increase the overall market for e-commerce through its unique value proposition of providing easily personalized local goods and services anytime and anywhere. An interesting study by IDC states that by the year 2006, e-commerce spending is predicted to grow to $38 billion, representing a compound annual growth rate (CAGR) of 32.3 percent through 2006. Also, according to a recent study by Forrester, three-quarters of US households will have at least one "Internet-capable" device by 2005. The PC will still be the primary computing device, but other products—including wireless Web devices—will find a home with North American consumers. According to market research firm Strategy Analytics, the global market for m-Commerce is expected to reach $200 billion by 2004. The number of Internet subscribers will grow from 90 million in 2003 to over 200 million by 2007. In addition, Strategy Analytics expects the U.S. cellular market to grow to 8.8 percent in 2003. Subscribers will increase from to 153.2 million in 2003 and 202 million at the end 2008, with penetration reaching 68.5 percent.

Those numbers represent a significant increase in the wireless user community, and thus an even more significant demand for wireless services. Figure 2.8 takes a look at how m-Commerce is poised to take off and describes how it will have a profound impact on business models and lifestyles.

	Key Trends	Implications
Industry Dynamics	• Competition will be fierce. • Margins will fall. • New m-Commerce applications offer new revenue streams.	The high cost of 3G licenses, low margins and increased competition will drive consolidation among mobile operators in Europe. m-Commerce will turn many operators into banks. Customers will charge transactions to their mobile bills. Prepaid accounts will unlock e-Commerce for the large numbers who cannot hold credit cards.
Technology	• Technology will drive m-Commerce development. • Beware of "hype."	m-Commerce will only boom when technology supports high-quality functionality. The infrastructure required to deliver these improvements will be expensive, requiring significant changes to hardware in place.
Customer Relationships	• Customer ownership is key. • Personalization and convenience will optimize value.	Customer interaction will shift from administrative tasks to relationship building. Mobile Internet will allow to service customers from a distance without the cost of human agents or the hassle of voice response. Contact with the customer will focus on improving the customer relationship, ensuring that they will repeat the experience.
Commerce	• Explosive growth is predicted. • Many new business and consumer applications are foreseen. • Will be driven by personalized and localized offerings.	Phones will become keys, tickets, and coupons. For example, after booking a holiday, the mobile device will store the flight tickets and hotel reservation, and will serve as the key to the customers room. Currently, m-Commerce focuses on human applications. In the future, m-Commerce will enable intelligent devices, e.g., Cars will report their performance and malfunction details to manufacturers to influence design improvements, highlight problems and schedule services.

Figure 2.8 Mobile commerce trends.

Process Improvement Opportunities

Mobility offers new opportunities to dramatically improve business models and processes and will ultimately provide new, streamlined business processes that never would have existed if not for this new phenomenon. Figure 2.9 illustrates how processes will evolve to factor mobility.

Let's go through each of the prime tenets in Figure 2.9 (signified by the stars in the chart).

- **Extending Web to wireless.** The first phase in the evolutionary track is called extending Web to wireless. This is also known as "Webifying" or extending your existing Web presence. For the most part, business processes are minimally affected in this phase. The goal is to provide value-added services through mobility with minimal disruption to existing processes. An example might be creating a new company Web site accessible through WAP (Wireless Application Protocol) phones or Palm OS–based personal digital assistants (PDAs). Firms attain immediate value through realizing additional exposure and market presence, and customers realize value through additional services.

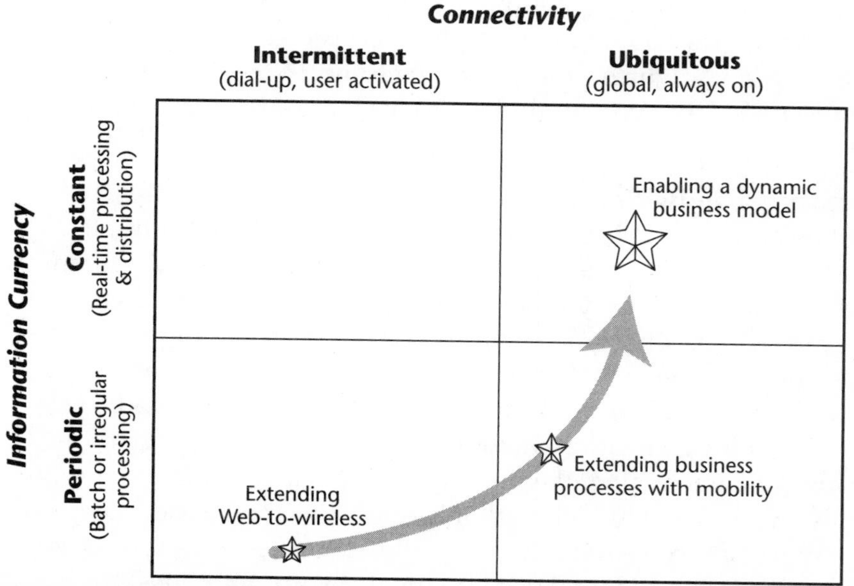

Figure 2.9 The evolution of mobile business models.

- **Extending business processes with mobility.** The next step in the evolution of mobility is to extend existing business processes while new processes are born through adaptive business models. New opportunities to streamline company business processes emerge and evolve to produce new revenue opportunities. One example is the way that mobility extends business processes through a supply chain optimization model. Real-time tracking and alert mechanisms provide supply chain monitors with the capability to monitor shipments and product line quality in ways that traditional business models were not capable of doing. New business processes emerge through these new mechanisms that ultimately shorten the supply chain cycle, thus minimizing error and maximizing efficiency and realizing the utmost customer satisfaction.
- **Enabling a dynamic business model.** The final phase in the mobility life cycle is one that has only been touched upon in today's world. The unique attributes of mobility will provide new and exciting ways of managing processes and allow for efficiencies never before attainable. The convergence of wireless technologies with existing business models will become the mobile "nirvana," or a fully dynamic business process.

Enabling Technologies

Mobility brings a whole new world of technology to our fingertips. Many of these technologies have already been discussed in this book to some extent. However, it's worth going through some of these enabling technologies and issues that you might be currently experiencing with mobile technology.

- **Increasing bandwidth.** The first enabler, and by chance also the most discussed in the mobility realm, is network bandwidth. If you've been reading the newspapers or magazine articles pertaining to wireless technology, you've probably come across the discussions on the dynamics of network bandwidth and wireless infrastructure. In recent years, one of the most heavily talked about issues is when network and telecommunications providers will launch and complete 3G (third generation) networks. Of course, the key factor in 3G versus 2.5G and analog networks is the capability to be "always on." In other words, not having to dial up each time you want to perform an activity or receive data. Additionally, mobile devices can access data at blazing broadband speeds similar to our wired desktops. At least that's the idea. New functionality, like streaming music and video, will open up new markets for consumer's enterprise customers. Most network providers have begun or are completing implementation of 3G services. We'll cover 3G and other bandwidth issues later in this book.

- **New devices.** The plethora of mobile devices available today often makes the buying decision-making process painful for customers. With each new device appearing to be better, faster, and more powerful than the last, how are we to make a choice? To put some of the different types of devices in perspective, let's go through the various categories of mobile devices. Basically, mobile devices come in four categories, those being:
  - **Unintelligent gadgets.** This includes devices such as sensors and RFID (Radio Frequency Identification) tags. Unintelligent gadgets typically contain little or no processing power. However, these types of devices can pack a great of functionality in a small device. A good example of RFID in use is vehicle-tracking tags. The idea here is that a small sticker-like RFID tag is placed on the inside windshield of the vehicle. The tag stores critical information about the car, the person driving the vehicle, or other data that can be used to identify the vehicle. Using radio transmitters, the car can be located, say in a parking lot at an airport, or if the vehicle was stolen. RFID tags can be used in many types of applications. To some extent, they are like mobile wallets; they can store credit card information, allowing purchases to be made at restaurants and any retail establishment willing to accept the technology.
  - **Cellular phones (mobile phones).** This device category is probably the most familiar to most of us. It includes voice technology along with WAP, Short Message Services (SMS), and wireless Internet capabilities. As 3G phones are finding their way into the market, these services are now becoming commonplace throughout the world.
  - **Smartphones.** These devices are still maturing in the mobility market space. They typically offer a combination of voice, data, and an operating system like Pocket PC or Palm OS. They provide multifunctional displays and usually appear as combination mobile phone and PDA, like the Orange SPV Windows Smartphone. Many of these devices also support the disconnected or

occasionally connected modes of operations described in the previous section. They are capable of supporting thick as well as thin clients, which allows them to operate with a wide range of applications.

- **Devices with operating systems.** Although it is technically true that smartphones have operating systems, and paging devices may have operating systems, the distinction we're drawing here is to separate devices that are primarily of the PDA and personal computer ancestry rather than that of the phone. Some of these operating systems include the Palm OS, EPOC, the Pocket PC, Linux and the various versions of Windows and Macintosh. For many of these devices, the native mode of access would be assumed to be over a wireless LAN rather than through the cellular data networks, although either or both is certainly a viable choice. This is a new point of distinction, because the smartphones, as seen today, are not capable of this type of connection.

- **Voice navigation.** Another technology enabler that's making waves is voice recognition applications. We already touched upon a scenario where voice recognition and Telematics combine to allow users access information while in their cars. Speech recognition can also offer significant opportunities in the call center and customer service space. Voice commands can be used to navigate through applications such as portals, voice mail, e-mail, and almost any personal calendar management system. More sophisticated applications can even be used in the field service and sales force automation space for communicating status, and so on.

- **Localization.** Finally, localization enablers are still in their infancy, but are making a strong presence, especially where the technology is more readily available. We mentioned some of the unique mobile applications in the entertainment space that give game players the ability to play games and to track other players using the GPS and mobile phones. As localization technology enablers mature, we'll see significant opportunities for advertising and customer service applications.

New devices and technologies are constantly arriving in the mobility space. Frankly, that's part of the fun of being in this business! At the same time, it generates a great deal of "noise" in the marketplace that produces confusion in our customers. As you'll see in Part II, a mobility strategy will be key to taking your first steps in this exciting world. As a final wrap-up to this chapter, let's go through some of the wireless trends that you should be aware of.

Wireless Industry Trends

If you've been following the wireless market space the last couple of years, you probably know that it's one of the most exciting and dynamic industries to hit the radar screen. There's never a dull moment when it comes to new developments and new products becoming available. With those kinds of dynamics, customer demands and expectations are constantly changing as well. Once a new product is released, a newer and more functional version is released by a competitor. From the consumer's perspective, the choices are never-ending. From the producer's perspective, competition is fierce.

The big issue really is to understand where the wireless bandwagon is headed and what we need to do as business owners and consumers. It's hard to imagine that overall revenue from mobile voice and data services is set to reach 130 billion euros by 2006, up from original estimations of 95 billion euros in 2001 by data researchers at Analysys. Assuming those estimations are even close to being accurate, the opportunity for wireless applications and services is staggering! To achieve this adoption rate, however, applications and functionality must be developed to improve the value proposition for potential business and consumer users. In both Europe and Japan, the "killer app" is SMS, which we'll discuss in Chapter 9. SMS provides the capability to send and receive text messages as well as the standard information services (news, stock quotes, and so on) over standard wireless phones. Short messaging service experienced exponential growth in Europe once it reached market penetration of 20 percent. Overall, more than 2 billion SMS messages are sent throughout Europe monthly, as compared to the 30 million people a week that use paging services for instant messaging in the United States. In short, wireless instant messaging may be the application that provides a foothold for the adoption of mobility in the United States. For a comparison on the growth of wireless in the United States, see Figure 2.10.

In the next decade, wireless devices will provide access to the Internet for more users than computers. Although new technologies will provide bandwidth at speeds equivalent to today's wireline connections, the user experience on the wireless Internet will still be significantly different from that on the wired Internet. More emphasis will be placed on the localization, personalization, timeliness, and convenience of information and transactions. These and other topics are covered in Chapter 9.

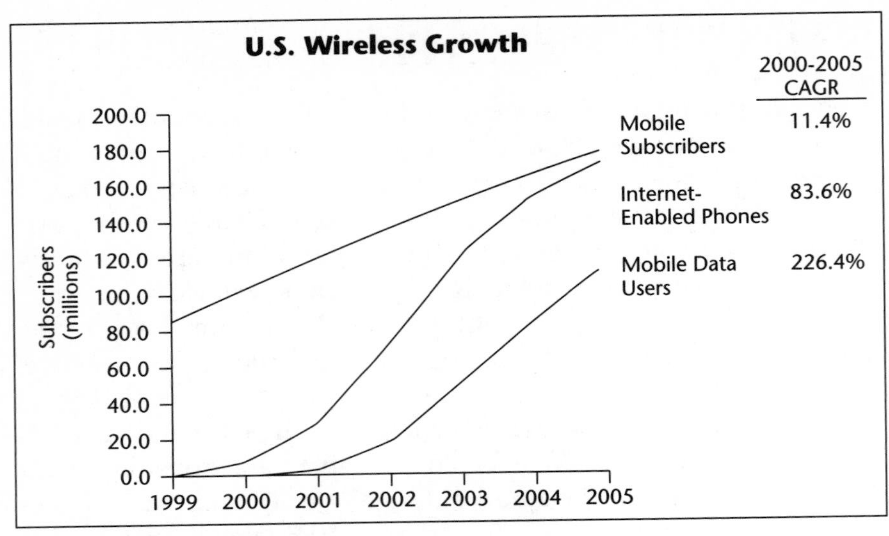

Source: Forrester Research

Figure 2.10 U.S. wireless growth.

Summary

As you've seen in this chapter, the business models around mobility are not too unlike what we've seen in the "e" world, or the e-business-driven environment. Companies usually take either a customer- or consumer-driven approach, or a business-driven approach. In other words, applications are built around business-to-customer models or business-to-business models. Examples of the consumer-driven mobile applications that are offering compelling value propositions are wireless gaming and location-based applications. Business drivers range from mobilizing the sales force and field service applications to executive dashboards and wireless local area networks such as WiFi. In the next chapter, we'll introduce the various mobile devices on the market and where they fit into your wireless strategy.

CHAPTER

3

Mobile Devices Showcase

Michael Welin-Berger

Introduction

This chapter describes the different alternatives you have when deploying mobile solutions. There is no "one size fits all." On the contrary, all user groups need to define their specific needs and equip themselves accordingly.

Our belief is that you will use many different devices to connect to the Internet. Here, it is necessary to clarify that we refer to the Internet as one possible communication channel (mostly wireless) that will link all devices to different services. Furthermore, there are no great differences between the devices you will use for professional and private purposes.

To see what these devices will look like, we need to address how we will use them. Here is a list of relatively short-term predictions concerning the ways in which we will use mobile devices. Its purpose is to help you understand how existing devices will develop, which is the main topic of this chapter.

- Your trusty companion will be a personal digital assistant (PDA). It will have the same size as today's Palm or Pocket PC, but the services it provides will be much better. Your PDA (or whatever the device will be called in the future) will also be your cordless hands-free telephone, or your mobile TV set. The reason the PDA will be larger than today's cell phones is simple: screen size. A big screen will be necessary if you want information presented to you graphically, which gives it another look and feel. The device will, however, virtually only be a screen. It will be considerably lighter than today's PDAs.

- In your wallet, you will always carry an extra telephone the size of a thick credit card. It will be used when you've forgotten your PDA in the car, or if you've chosen not to bring it along when taking your kids to the movies.
- Finally, and most frequently, a flat screen will be located on the wall in several rooms in your house (or at work). These screens will have a touchscreen functionality but you will also be able to use mobile keyboards. The flat screens will replace today's TVs, PCs, and stereos—and they'll be connected to your information base containing movies, music, e-mail, and any other information you've stored to be accessed at your convenience. Flat screens will also be the tool of choice when you don't need to be mobile—regardless of whether you're at the office or at home. The mobile version of flat screens will be similar to the Tablet PCs that have recently been promoted by Microsoft. But the connected screen of the future—some suppliers are already talking about the "Surf Board"—will be much lighter.

Predicting what mobile devices will look like in the future is interesting, but an even more interesting task is foreseeing the kinds of services we will use. The functionalities we will use in our devices will most likely be offshoots of services that are available to us today. Voice command is such a service. Certain cell phones already offer this functionality: You say the name of the person you want to call and the call is connected. This is a very basic example of connecting a command to a certain process. It will take some years to introduce more advanced services and commercial applications, such as voice navigation of the Internet, for large user groups.

Merging Computers and Phones

If you look at how PCs were equipped some years ago and compare them with today's Pocket PCs, you will see that they have the same performance. Pocket PCs typically have a 200-MHz processor and 64MB of memory, which was the standard for PCs some years ago. Now that color screens are available on these devices, a whole new set of opportunities for how to use these devices arises. The next step for handheld computers is to integrate a telephone functionality in the device—examples of this are already available on the market.

If we look at cell phones, we can see the emergence of two alternative paths: developing smaller phones or merging phones with computers. Cell phones are getting ever smaller, and carrying them around in our pockets will be as common as wearing a watch on our wrist. When it comes to merging cell phones with computers, this has already been achieved—the result is a so-called smartphone (or communicator). Smartphones are the telephone supplier's answer to the PDA—they bring simplified computer functionalities into the phone. The most common operating system is EPOC (also referred to as the Symbian OS), which gives you the capability to surf the Web, read your e-mail, and download other applications to your phone.

In the future, the differences between PDAs and smartphones will gradually diminish, and there will be no reason to distinguish between them. For the time being, however, the devices will still have different characteristics. Here are some examples of the differences that we see today:

- Weight—related to battery capacity
- Screen size and image quality (resolution)
- Performance available for installed applications

Laptop Computers

The laptop computer is the companion of many a mobile worker. It allows people to work wherever and whenever they want to. Most of the information they need resides in the laptop, and there is also the possibility of connecting to the Internet to gain access to e-mail and other information available online. Primarily due to their large screens and full-size keyboards, laptops are great for data-input heavy activities, such as the creation of documents, spreadsheets, or presentations. They do, however, have some drawbacks that reduce their usefulness to mobile workers. The computers' large size and often quite substantial weight still make them relatively cumbersome devices, especially to those users who need continuous access to a computing device while on the move. Although the laptop computer will remain the device of choice for many mobile workers who will be stationary at various intervals during the day, they will be less useful for individuals who need computing power while walking around—as is the case for a mobile field technician or a doctor making hospital rounds.

Laptops will continue to evolve, with the next generation evolving into Tablet PCs. Tablet PCs are similar to laptops but there is one major difference: the screen is placed on the outside of the device rather than on the inside (you don't need to flip up the lid to access the screen). Tablet PCs have a touch-screen keyboard, which does make them less useful when writing extensive documents. Nevertheless, the capability to communicate and to have access to all the functionalities of a PC will be appealing to many users.

Some mobile workers will move from using a laptop computer to working with a highly mobile PDA that is linked to a stationary PC at the office. However, most "true" laptop warriors will continue to prefer working with fully functioning, large memory and processing power–bearing laptop computers that have bigger screens and keyboards than their lightweight PDAs.

Personal Digital Assistants (PDAs)

A PDA is a small computer with a screen that is substantially larger than that of a cell phone. PDAs are the most fascinating devices when it comes to hardware development over the next few years. We will see fierce competition in this area, which will give rise to lighter PDAs that have higher resolution, increased memory, longer battery life, and more processing power. Another challenge that PDA manufacturers will face is the integration of phone functionality into the PDA. This certainly is a daunting task, but cell phone and PDA manufacturers will probably combine their efforts to find a solution. There is no question that PDA suppliers will meet the need for data communication—what's more uncertain is if they will create a satisfactory telephone functionality. As of the writing of this book, there are several efforts already underway, but no clear winner is apparent.

PDA manufacturers are one step ahead of smartphone manufacturers—they have advanced know-how for building devices that are similar to PCs. The greatest challenge will be to create a relatively small device with functionalities that are virtually identical to those of a PC. In essence, the graphical user interface will be combined with an operating system that enables databases and rich applications. This is quite similar to building applications for a PC.

There are currently two main PDA alternatives on the market: the Pocket PC and the Palm Pilot. The Ipaq from Compaq was a great success on the Pocket PC market, closely followed by HP's Jornada. After the merger between HP and Compaq, the new HP (www.hp.com) will be a strong player in this market. The question now is which company will create the next generation of devices on existing platforms. Although other suppliers have launched PDAs running the Pocket PC operating system, they have yet to reach HP's popularity.

Palm (www.palm.com) has an advantage over its competitors thanks to the company's market penetration (although the company is losing market share, it still sells the most devices). Palm manufactures its own devices, but it now faces competition from Handspring (www.handspring.com), a company that actually incorporates Palm's OS in its devices. Here, Palm finds itself in a similar battle to the one Apple (www.apple.com) fought concerning its Macintosh OS. In that case, Apple was the first to create a graphical user interface, yet after a successful launch had to face Microsoft when the company entered the market.

A third PDA alternative has just recently been introduced in the form of devices built on different versions of Unix. Linux is the most well-known platform that will be available on different devices, such as the Sharp Zaurus. This PDA is positioned to compete against both Pocket PCs and Palms, but it is too early to declare a winner.

Palm Family (Palm OS)

Palm currently enjoys the biggest share of the PDA market. The usage of Palm's devices is mainly limited to PIM (Personal Information Management), a functionality that enables you to access your calendar, e-mail, address book, and to-do lists. Even though most users synchronize the Palm with their PC by placing the device in a cradle that connects the device to the desktop (or laptop) computer, they do have the possibility of connecting wirelessly in order to exchange e-mail and update their calendar.

The growing number of enterprise applications being rolled out today presents a challenge to Palm, because they often require capacity for local databases and local applications, both of which demand high performance from the handheld. This task reveals the weakness of the Palm OS, namely its lack of capacity. Until now, the focus has been on creating a device that, when used normally, only requires battery recharging once a week or once every other week. The goal of low power consumption has affected the Palm's screen. Although it is less power consuming than most other Pocket PC devices, it has the poorest screen quality on the market. The power restrictions also affect processor performance negatively, but this doesn't apply to PIM since the applications are adjusted to the processor's speed. When building a new application containing information from ERP (enterprise resource planning) systems, however, the limited performance of Palm devices is a restriction.

When critics of the Palm address the benefits of using a Pocket PC, their main argument is that the Palm hardware is too limited to enable the functionalities we want. On the other hand, Palm fans claim that high performance isn't necessary if the application is built correctly from the outset. We will avoid choosing sides in this battle, but it is important to note that both camps have convincing arguments. Palm's greatest advantage is that, due to its market share, people are getting used to working with the handheld. They are, in other words, used to the Palm's metaphors and know how to navigate on the device. Similarly, one of Microsoft's advantages is the resemblance in look and feel to Windows. Although Palm and Pocket PC are similar in some respects, each device has a distinct "look and feel" to it that people become accustomed to and like.

Pocket PCs (Windows CE)

The Pocket PC is a combination of Windows CE and an added functionality, which is similar to that of Microsoft Office on full-blown desktop and laptop computers. Pocket PCs contain, among other applications, e-mail, a calendar, an Internet browser, Word, and Excel. The name confusion between Pocket PC and Windows CE really doesn't matter, and Microsoft has been extremely successful in promoting the Pocket PC as the platform for future mobile devices. The most encouraging fact concerning sales is that the Pocket PC seems to be the platform of choice in the enterprise market when it comes to connecting devices to back-office systems. In the personal domain, that is, for people who want a PDA for personal use, the competition from Palm is stiffer.

Pocket PCs will obtain a substantial share of the enterprise market due to the fact that the technology for building applications is based on the familiar Microsoft platform. The platform of the Pocket PC presents great benefits to companies that already use the Microsoft architecture, because this enables standardized connections from the device to the enterprise applications.

When compared to a Palm Pilot, Pocket PC devices are more powerful in terms of memory and power. The screen in the Pocket PC is generally very bright, and you can easily read the content even in bad light. The downside of bright screens is the power consumption—you usually need to recharge the battery after only one day of frequent usage. This problem becomes even more apparent when you connect by way of WLAN or GPRS (General Packet Radio Service), for example, which requires even more power, resulting in fewer hours of operation before recharging. Thus, the biggest hurdle the device manufacturers have to overcome is creating a device with substantially longer battery capacity without increasing its weight.

Two-Way Pagers

While Europe focused on SMS and connecting the PC via GSM (Global System for Mobile Communications), the pager has been the alternative of choice in the United States Recently, the pager has become increasingly similar to the PDA. The difference is that pagers focus on services like e-mail and the calendar rather than on the use of an Internet connection and a browser. When we look at recent versions of two-way pagers, we find two devices that have increased their functionality along with the introduction of new PDAs. The two most common examples of two-way pagers are the BlackBerry from RIM (www.rim.com) and the Timeport from Motorola (www.motorola.com). The reason for the expected growth for these devices is twofold: their functional similarities to the Palm, and their capability to connect even in remote locations.

To understand how the BlackBerry and the Timeport work, it is useful to look at how the network communication works. The pager was initially built on Mobitex network, but has now been extended to several different networks. The technical description is as follows:

> *Mobitex is a packet-switched, narrowband PCS network, designed for wide-area wireless data communications.*

In plain English, this means that the network provides slow data communication with good coverage. This type of communication is used for trucks and all sorts of equipment that is reliant upon good coverage. Mobitex has existed in both Europe and the United States for more than 10 years, but its breakthrough for personal use was primarily due to the pager. In the late 1990s, the BlackBerry made it even more widespread with its ease of use and growing network coverage. Two-way pagers are not meant for Web browsing; they are more appropriate for services such as maintaining calendars, keeping lists of contacts, and exchanging e-mail via the device. Because the functionality of viewing attachments is rather limited and the connection speed is slow, there are clear restrictions on the size of e-mails.

One of the biggest differences between two-way pagers, Palms, and Pocket PCs is the keyboard. Even though the keys are small, pagers actually have a real keyboard. This makes for a considerable advantage when composing long e-mails.

Today, RIM, Motorola, and other two-way pager manufacturers face the challenge of adapting their devices to the new networks. BlackBerry and Timeport have already introduced devices that use GPRS communication, which is the data communication format used in GSM networks. This will open a new market where the interest for these kinds of devices is high. They now have to compete with smartphones and PDAs. The biggest need for adapting to new networks emerges from the speed that becomes available; Mobitex is simply too slow.

Cellular Phones

We have seen an explosion in cell phone usage. In some countries in southern Europe (for example, France and Portugal), mobile penetration is higher than that of fixed-line telephony. The biggest surprise has been Europe and Japan, where the youth market has become an important user group.

The usage has been higher in Europe and Asia than in the United States, but the latter is catching up as a result of new networks that enable better coverage and lower fees. The technical evolution has gone from analog networks like NMT (Nordic Mobile Telephony) and CDMA (Code Division Multiple Access) to digital networks such as GSM and TDMA (Time Division Multiple Access) that enable data communication with much higher bandwidth (greater speed).

The digital networks enabled WAP (Wireless Application Protocol), which was expected to reach high penetration levels after the millennium shift. WAP is a text-based protocol, which does not enable the use of graphics in its applications. However, the expectations were not met. Despite the fact that most mobile phones in Europe are WAP enabled, very few people actually use existing WAP services. The reason for this failure is the poor text-based interface in combination with a relatively complex connection procedure. Furthermore, end-user pricing associated with using WAP services also acted as a deterrent that prevented widespread adoption of this service.

This makes smartphone users the only group that actually uses WAP—they've gotten accustomed to reading e-mail on their smartphones. The smartphone is a cell phone with some PDA functionality, which often uses WAP for communicating with services residing on the Internet (mail and calendar services).

Smartphones

Smartphones are what telephone suppliers want us to use as our personal companion. Their concept is built on implementing computer functionalities in cell phones. The basic functionality provided by today's smartphones is the PIM service (e-mail, calendar, and to-do lists). The main difference between a smartphone and a PDA—apart from the phone capability—is that smartphones have poor capability when it comes to incorporating additional applications. In short, smartphones are still more like phones than small computers. This can clearly be seen in their screens, which have relatively poor functionality. With smartphones, you predominantly make use of a browser when you're connected to the Internet.

There is a joint initiative to provide a common platform (EPOC OS) for smartphones. The name of the initiative is Symbian and its members are, among others, Psion, Ericsson, Nokia, and Motorola. All members already have devices for EPOC; what has been missing so far is common initiatives regarding security and other similar features. Examples of devices already available are the Nokia Communicator and the Ericsson P800. The next step is to provide good Java compatibility for the EPOC OS. This means that the way in which Java applications are downloaded and executed will be the same in all devices. This is a complicated task due to the differences between the devices (screen size is one such difference).

Microsoft has now launched their smartphone, and it will probably become a success in the market due to the user experience, which is very similar to the traditional Windows interface. We have not yet seen larger rollouts of these devices, but as a repository for all PIM, e-mail and with a limited online functionality, we expect rapid growth in market share. The biggest hurdle will be to educate the user in inserting information with a phone keyboard, as is done when sending traditional SMS with a phone.

Imode Phones

Imode is one of the most fascinating initiatives the mobile world has seen so far. NTT DoCoMo in Japan created the Imode service by first defining the entire architecture and functionality in the device and then delivering the specification to a number of device manufacturers. NTT DoCoMo went on to assume responsibility for the launch of the services, including a payment model in which they, as operator, take 9 percent of the revenue. By providing the content providers with a business model, NTT DoCoMo created all the necessary components for the success. In this case, one operator on the market single-handedly decided how an entire concept should work, thereby eliminating many of the problems that usually stem from a lack of standards for mobile services. This differs significantly from how operators have acted in Europe and the United States, which can partly be explained by the fact that these markets are more fragmented than Japan's. *Fragmentation*—less market share per operator—makes it more difficult to establish a proprietary standard.

The functionality in the first version of Imode was not very impressive. It offered relatively low bandwidth (speed) for sending images and ring tones. Imode nevertheless became a success, mainly because the services it provided found a user group that was prepared to pay for them. That user group was the youth market, teenagers between 12 and 25, which explains the focus on entertainment, horoscopes, and similar content. We will, in coming versions, see services that have much better functionality and bandwidth.

Imode is now launching its next generation of services, which builds on downloadable Java applications that enable more complex functionality. We describe how this works in detail in the next section. The key finding from the Imode example is the need for a strong process owner in situations where a technology lacks established standards.

Phones Using Java

Java is a programming language that has grown rapidly in usage in recent years. It is available on all computer platforms and is now entering the world of mobile devices. Java was introduced to the market by Sun Microsystems, but it received support early on from the huge Unix community around the world. Some of its characteristics are that it is object oriented and has the capability to send applications to a device. These applications can then be executed in a PC or on a handheld.

Our earlier discussion concerning thin and thick clients (that is, whether an application resides in a server or on a device) now has a third option. This option is the use of so-called Java applets, or codes that are sent to a device and then executed on the device. Let's look at an example.

You want to connect to your bank account to check your balance by using your PDA. After the authentication process, you choose the option "check my savings account balance." The corresponding application is then downloaded to your PDA and you can check your balance as well as all your latest transactions.

There are two major benefits to this approach. First, you will experience much faster communication because most commands are merely executed on the PDA—all the information doesn't have to be fetched from the central bank application. Downloading the application will not take a long time because it is possible to build the application in such a way that it is transferred from the bank to your PDA in just a couple of seconds. The second major benefit is that the application can be stored on the PDA once it is downloaded. This eliminates the need to download it again the next time you wirelessly connect to the bank, say the following week. Instead, all you have to do is run the saved application to gain access to your latest, updated savings account balance.

Java has very high expectations from the mobile industry. It remains to be seen if these expectations will be met.

Handheld Gaming Platforms

The gaming industry is focusing on existing phones and PDAs instead of providing specific devices for games connected to the Internet.

There are some companies, such as Cybiko Inc., that manufacture small, handheld gaming devices which can be connected wirelessly to a group of devices that are within 300 feet of each other. This is an example of the kind of devices that we assume will populate the market in a couple of years. Cybiko will, however, require direct Internet access instead of mere local connection. Nintendo—the manufacturer of the GameBoy device—is expected to launch something in this area, but there seems to be more caution in the decision-making process today then there was in the late 1990s.

When we look at the alternatives and compare them with the development of stationary gaming platforms, we expect to see still more games for existing PDAs and smartphones prior to the launch of wireless gaming devices on the market.

Future Devices

Five years from now, we will be using new kinds of equipment and entering information in a different way. Glasses (or goggles) will replace today's screens, and we will be using our voices and virtual keyboards to enter information and navigate in applications. Let's take a closer look at some of these potential functionalities.

The virtual keyboard already exists, and it has been shown in action. You put straps over the palms of your hands. The keyboard then recognizes how you move your muscles and can subsequently identify which keys you would be using on a QWERTY keyboard. The clever part is that there is no keyboard—you are simply tapping away into thin air. This is a good solution for situations when you normally can't use a traditional keyboard (for example, when you're sitting on a bus). Similarly, there is a company that has developed tiny microchips, the size of a pinhead, that can be glued to your fingernails. Again, without using an actual keyboard, you can "type in the air" while the chips record your fingers' movements and translate these into typed letters.

Glasses that show a screen are also already available; they have been used on oil platforms when technicians are mending equipment in bad weather conditions. They have evolved from the 3-D glasses used in virtual-reality environments. They do, however, need to become smaller and lighter than they are today.

Now, let's combine the aforementioned virtual keyboard and glasses with a powerful PC that is attached to your belt. The PC will weigh only a few ounces and be the size of today's cell phone.

This scenario might sound far off today. But try to imagine what would have happened if you had heard about the performance of today's PDAs in 1997. You probably wouldn't have believed a word!

Summary

As stated in the introduction to this chapter, our belief is that you will use many different devices to connect to the Internet. When we describe the different alternatives available today, you will probably find devices that will suite your usage model well. However, the quick evolution among devices requires that you update yourself about such devices before making any major purchase, thereby ensuring that you get the best available equipment.

Also remember that no device is better than the support it gives to the users, which means that there needs to be a thorough design of the usage models to ensure that you choose the appropriate equipment.

CHAPTER

4

Wireless Networks Overview

Adam Kornak

Introduction

An area that we've spent little time on in this book up to now is the wireless network. This book is, after all, intended to be a guide for mobile business and strategy, but it should not exclude an understanding of the technology that makes it all happen. It's important to understand the technical and physical infrastructure of mobile systems to better understand how the business applications are implemented. Therefore, this chapter describes the evolution of wireless technology throughout the generations. Wireless networks have gone through many evolutionary changes, from the early analog bag cellular phones to today's fully digital 3G (third generation) smartphones. You'll find that wireless technology is usually broken down into generations. The generations we'll review in this chapter are:

- First generation (1G): Analog
- Second generation (2G): CDMA, TDMA, and GSM
- "Two and a half" generation (2.5G): GPRS and EDGE
- Third generation (3G): CDMA2000, wCDMA, and UMTS
- Fourth generation (4G): Voice-to-text and OFDM

This chapter is still written for the business savvy but with a technology twist to help our readers understand why they might want to use certain technologies for business applications. So, we won't go into the nuts and bolts of the wireless network. But, if you are interested in learning more about the deeper architectural concepts of wireless networking, check out our book *The CGE&Y Guide to Wireless Enterprise Application Architecture*, available through John Wiley & Sons. As for this chapter, we'll continue with a discussion on an area that has made enormous advancements in the last couple of years, wireless LANs. You probably understand that it's almost impossible to forget the topic of wireless security and what any business strategist should be aware of in deciding on a wireless network. We've left the topic of securing a wireless infrastructure for the end of this chapter. Right now, let's discuss how the different generations of wireless networks have defined mobility and how applications in this exciting space have evolved through the years.

The Generations of Wireless

It's hard to imagine that in some parts of the world, wireless devices and their corresponding networks are nearly as powerful and as fast as the desktop PC. That excitement is driving service providers and software developers around the globe to design mobile applications that will enable business users to perform functions that were never even remotely considered and to leave their private cubicles and offices. From the simplest function of sending and receiving e-mail messages to more complex functions for monitoring the supply chain processes, mobility has come a long way since the days of the first cell phone. With that in mind, let's continue our discussion on the wireless evolution to obtain an understanding of the first generations of wireless networks.

First Generation: Analog

The first generation analog cell phone was a rather bulky piece of equipment that was typically owned by only a fortunate few who could afford the device, or by individuals who didn't mind carrying around a brick-sized phone. Quite often, the phone came with its own bag and antennae that protruded out of its pockets or was accessible through a briefcase. This invention nonetheless was the exciting beginning to cellular technology for the consumer and business marketplace. The concept of "mobile" devices for voice communication was truly nothing new at the time (early 1980s). However, the important technology of leveraging a distinct network of "cells" to enable wireless voice communication was a revolution in the way people would communicate in the future. Without getting too technical, "cell" technology works by assigning multiple base stations to a subset of users, thereby increasing the system's capacity exponentially, while reducing power requirements for the user terminals, in this case the analog cell phones. Analog networks only have the ability to carry your voice, not data. So, it's impossible to send an e-mail or any other type of data element that requires the movement of digital information.

Analog networks still exist in most wireless network infrastructures, but are being quickly phased out by newer and more efficient digital networks. The reason that service providers must keep analog around is that the combination of unlike digital networks cannot readily communicate and interchange information directly with each other. In other words, AT&T's Digital TDMA (Time Division Multiple Access) network cannot communicate directly with Sprint's CDMA (Code Division Multiple Access) network. We'll talk about the differences between these digital networks in the next section on second generation wireless. Suffice it to say that analog cell networks were the first step to a world of mobile voice communication to be accepted by the mass-market consumer marketplace. Soon after, digital networks made their way into the hands of the consumer with the next generation of cellular technology.

Second Generation: CDMA, TDMA, and GSM

The second generation (2G) of wireless and mobile phone technology gave users the capability to send and receive data in a digital format. In technical terms, it's a way of encoding analog information into digital data using a binary language of 0s and 1s. Digital technology offers many benefits over analog by offering better service to customers (a service operator can fit more information in a transmission), much improved clarity of sound (during voice conversations), higher security, and access to future generation features. At the same time, one of the negative aspects of second generation digital is that three distinctly different digital networks exist that don't allow for interchangeable communication. As was mentioned previously, this is one of the reasons why analog networks still exist today. The three primary second generation digital networks are covered in the following sections.

CDMA

CDMA (Code Division Multiple Access) is a digital cellular technology that uses spread-spectrum techniques. Unlike competing systems, such as GSM, that use TDMA, CDMA does not assign a specific frequency to each user. Instead, every channel uses the full available spectrum. Individual conversations are encoded with a pseudorandom digital sequence. CDMA is actually a military technology first used during World War II by the English allies to foil German attempts at jamming transmissions. Interestingly enough, the allies transmitted over several frequencies, not one, making it difficult for the Germans to pick up the complete signal. Because Qualcomm Inc. created communications chips for CDMA technology, it was privy to the classified information. Eventually as more of the information became public, Qualcomm claimed patents on the technology and became the first to commercialize it. Figure 4.1 depicts a graphical perspective of how CDMA functions in conversations.

The architecture of CDMA is such that multiple conversations are transpiring at the same time, sharing the same frequency as other CDMA conversations. The CDMA systems decipher each of the conversations so that each listener understands whom he or she is listening to. In Figure 4.1, imagine that all of the people in the room are all multilingual, and only the people talking to each other understand what the other person is saying, because they are speaking the same language. Advantages of CDMA over analog systems include:

- Capacity gains of 8 to 10 times that of AMPS analog systems
- Improved call quality, with better and more consistent sound as compared to AMPS systems
- Simplified system planning through the use of the same frequency in every sector of every cell
- Enhanced privacy through the spreading of voice signals
- Improved coverage characteristics, allowing for fewer cell sites
- Increased talk time for portables

Source: Ericsson

Figure 4.1 CDMA cell networks.

TDMA

TDMA (Time Division Multiple Access) was released in 1984. It uses the frequency bands available to the wireless network and divides them into time slots, with each phone user having access to one time slot at regular intervals. TDMA exists in North America at both the 800 MHz and 1900 MHz bands. Major U.S. carriers using TDMA are AT&T Wireless Services, BellSouth, and Southwestern Bell. A graphical depiction of TDMA is shown in Figure 4.2. As you can see, the TDMA architecture works in a "timeslot" format. In other words, one person speaks, and another is listening. For another person to speak, a timeslot (channel) must open up. Only one subscriber is assigned a channel at one time, and no other subscriber can access that same channel until the call is ended. As you can imagine, that makes for a great deal of channels. Advantages of TDMA include:

- Unlike spread-spectrum techniques, which can suffer from interference among the users all of whom are on the same frequency band and transmitting at the same time, TDMA's technology, which separates users in time, ensures that they will not experience interference from other simultaneous transmissions.
- TDMA provides the user with extended battery life and talk time because the mobile is only transmitting a portion of the time (from 1/3 to 1/10) of the time during conversations.
- TDMA installations offer substantial savings in base-station equipment, space, and maintenance, an important factor as cell sizes grow ever smaller.

- TDMA is the most cost-effective technology for upgrading a current analog system to digital.
- TDMA is the only technology that offers an efficient utilization of hierarchical cell structures (HCSs) offering picocells, microcells, and macrocells. HCSs allow coverage for the system to be tailored to support specific traffic and service needs. By using this approach, system capacities of more than 40 times AMPS can be achieved in a cost-efficient way.
- Because of its inherent compatibility with analog systems, TDMA allows service compatibility with dual-mode handsets.

GSM

GSM (Global System for Mobile Communications) is actually based on an improved version of TDMA technology.

> *In 1982, the Conference of European Posts and Telecommunications (CEPT) began the process of creating a digital cellular standard that would allow users to roam from country to country in Europe. By 1987, the GSM standard was created based on a hybrid of FDMA (analog) and TDMA (digital) technologies. GSM engineers decided to use wider 200 kHz channels instead of the 30 kHz channels that TDMA used, and instead of having only three slots like TDMA, GSM channels had eight slots. This allowed for fast bit rates and more natural-sounding voice-compression algorithms. GSM is currently the only one of the three technologies that provide data services such as e-mail, fax, Internet browsing, and intranet/LAN wireless access, and it's also the only service that permits users to place a call from either North America or Europe. The GSM standard was accepted in the United States in 1995. GSM-1900 cellular systems have been operating in the US since 1996, with the first network being in the Washington, D.C. area. Major carriers of GSM 1900 include Omnipoint, Pacific Bell, BellSouth, Sprint Spectrum, Microcell, Western Wireless, Powertel and Aerial.*

—From "Selling the Cell Phone/PCS Technology," an article by Mary Bellis

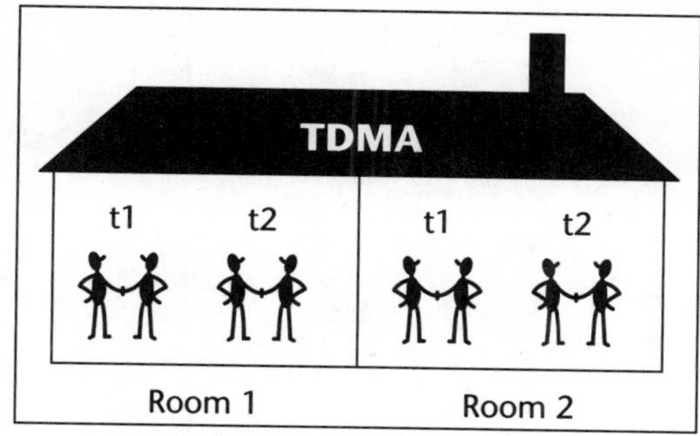

Source: Ericsson

Figure 4.2 TDMA cell networks.

2.5 Generation: GPRS, EDGE, and CDMA 2000

The next generation, called 2.5G, is actually more of an intermediate solution to third generation networks. Because third generation networks require a complete overhaul of the wireless network, the expense to complete the implementation is very high. At the writing of this book, most network providers are beginning to implement 3G and most are finalizing the process of taking the 2.5G plunge. The big question is what is this in-between generation and what does it mean from a business standpoint for wireless application implementation. As it sounds, the two and a half generation is intended to be a vast improvement in speed and services, but not to the extent of a full-blown 3G implementation. The additional functions provided by a 2.5G network are the following:

- Speed of data access (see Figure 4.3)
- Identification of location of the wireless device
- Ability to access customized information based upon location
- Ability to store information such as addresses and credit card numbers within personal profiles
- Ability to facilitate mobile online shopping
- Full mobility on the Internet
- Ability to provide business users with access to intranets

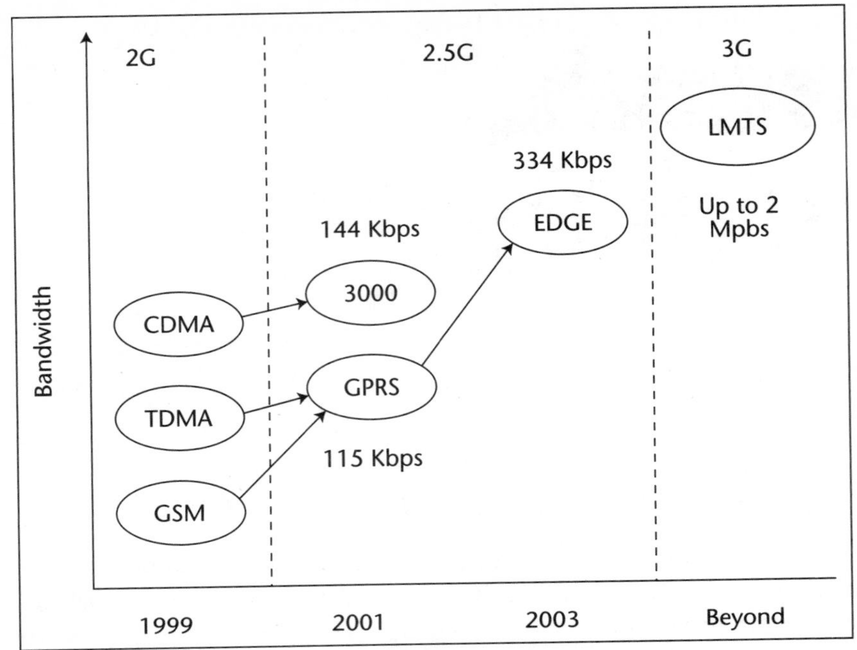

Source: Forrester Research, *The Dawn of Mobile Ecommerce*, CGEY Analysis

Figure 4.3 Evolution of wireless.

The technologies in the 2.5G space include the following:

- **GPRS (General Packet Radio Services).** Enables true "always-on" capability in the wireless network. Typically, in today's cell network, a phone call must be initiated to connect to a network, similarly to a modem dialing for service to an Internet service provider. By the same token, "always-on" can be compared with broadband-wired connections such as DSL (Digital Subscriber Line) or T1 lines and faster connections. However, GPRS only enables speeds in the range of 115 Kbps.
- **EDGE (Enhanced Data Rate for GSM Evolution).** Quite simply defined as a faster version of the GSM wireless service. EDGE technology enables data to be delivered at rates up to 384 Kbps on broadband connections. The standard is based on the GSM standard and uses TDMA multiplexing technology. In essence, the EDGE may enable higher functionality such as the downloading of music and videos over mobile devices.

- **CDMA 2000 (Code Division Multiple Access 2000).** The CDMA 2000 service is essentially a migration or upgrade of the CDMA standard discussed in the second generation section. CDMA 2000 is also an "always-on" technology that offers transmission speeds around 100 Kbps.

The next section turns to a discussion on the next generation of wireless, called 3G, or third generation, wireless.

Third Generation: wCDMA, UMTS, and iMode

The third generation of wireless industry technology represents the next major upgrade in functionality. Examples of 3G technology include the Universal Mobile Telecommunication System (UMTS) and wideband-CDMA (W-CDMA). 3G offers the ability to receive wired Internet features and services at ISDN-like speeds over a mobile handset. Unfortunately, 3G technology requires a complete overhaul of existing cellular networks. As a result, the immense capital investment involved requires a detailed analysis to determine whether the additional benefits justify upgrading 2.5G networks. Additionally, 3G adoption in the United States will be slowed by the competing standards and pricing schemes that charge wireless users for both incoming and outgoing calls.

Japan and Europe are currently leading the charge for the adoption of the next generation 3G data networks. Third generation networks are expected to arrive in Japan first, where capacity constraints on the existing cellular infrastructure necessitate an overhaul. Meanwhile, the European Union and national regulators have set a European 3G network implementation goal of 2004. With the United States being so far behind Japan and Europe, some analysts are predicting that fourth generation wireless technology could make third generation services obsolete before the 3G bandwagon even gets rolling.

Fourth Generation Wireless—What's Next?

While most of us are still thinking about when 3G will arrive in our neighborhoods, wireless network providers are already planning their launch for the fourth generation of wireless. 4G systems aren't planned for launch until 2010, although some providers have even insisted on a 2006 launch. The speed of 4G networks promises to be in the range of 100 Mbps and higher. According to the Fourth-Generation Mobile Forum, companies will have invested more than $30 billion in 4G by the end of 2003. That's a huge investment when companies are still trying to determine when 3G will be here and what it will offer. In fact, part of the reason that 4G technologies were and are being developed is due to the lag in the implementation of 3G networks. Although it's difficult to fully determine what applications will be offered by 4G, it's sure to provide a large number of opportunities such as real-time video and virtually any real-time application. In the next section, we'll discuss one of the hottest topics in the wireless industry today, the wireless local area network.

WLANs and PANs

The popularity of wireless LANs continues to grow as the mobile workforce demands access to corporate data while on the go. In fact, "more than 21 million Americans will be using public wireless local area networks (WLANs) in 2007, attracted by the cheap and superfast remote Internet access provided in airports, shopping malls, coffee bars, and hotels," according to a report by Analysys. Typically, end-users access wireless LANs through laptops or PDAs in offices, airports, hotels, or even their own homes. "The appeal of these services means that the number of hotspot locations in the U.S. will grow from 3,700 this year to 41,000 by 2007," said Monica Paolini, a coauthor of the report. "This will, in turn, generate over $3 billion in service revenues." The development of the WLAN market has been spurred on by the industry-wide adoption of a common technical platform based on the IEEE 802.11 standard, which we'll discuss shortly.

IEEE 802.11

In the wireless LAN market, the main technology driver has been the IEEE (Institute for Electrical and Electronics Engineers) 802.11 standard. The IEEE standards body has approved or is in the process of approving and standardizing three wireless LAN standards: 802.11b, 802.11a, & 802.11g. The primary differences among the standards can be easily described as follows:

- **IEEE 802.11b Standard.** The 802.11b standard operates at a maximum of 11 Mbps bandwidth range. 802.11b is in rather high demand at the current time as 2.5G and 3G networks are being implemented. 802.11b is absolutely not considered a replacement for these latter generation networks, but at the same time is an admirable substitute.

Many of the areas in which 802.11b, also called WiFi, is being marketed are hotels, airport terminals, businesses, coffee shops, and homes. All of the previous locations for leveraging this technology are turning out to be very promising because they are also locations at which 3G phones and devices would be used. Also, because laptops and PDAs are extremely rich in functionality and Internet capability, 802.11b is causing quite a stir in the marketplace.

- **IEEE 802.11a Standard.** IEEE 802.11a standardizes wireless devices operating at 5 GHz. "The specification for 5GHz was developed prior to the changes in 802.11b at 2.4GHz, and hence the odd numbering sequence. Bandwidth will scale from 6 to 54 Mbps. Future implementations of 100 Mbps are still being considered. 802.11a implements dynamic frequency selection, transmit power control and inter-access point protocol, features found in WAN cellular systems" (Gartner Group). Transmit power control enables a wide variety of cell sizes, permitting some applications to enjoy greater frequency reuse (which is important in high-density applications, such as stock exchanges). 801.11a employs between 40 mW and 200 mW of transmit power, compared with the 100 mW nominally used for 2.4 GHz. Despite technical enhancements that attempt to ensure transmission characteristics similar to those for 2.4 GHz, 50 percent more access points will be required to maintain a comparable footprint to a 2.4 GHz installation (0.6 probability).
- **IEEE 802.11g Standard.** The 802.11g working group is examining an upgrade to the 802.11b standard operating at 2.4 GHz in order to offer higher-speed services. Two alternatives are: (1) use of direct sequence to 20+ Mbps, and (2) use of orthogonal frequency division multiplexing (OFDM) to 54 Mbps. The 802.11g committee is split on which direction to take. Because of the emergence of 802.11a and the fact that a direct sequence improvement would permit users to upgrade, the first option will likely be the appropriate choice (Gartner Group).

Bluetooth

Bluetooth is a short-range wireless technology that connects electronic devices, including cell phones, printers, digital cameras, and handheld computers. Bluetooth is designed to exchange data at speeds up to 720 Kbps and at ranges up to 10 meters.

The technology of Bluetooth could take the computing world by storm. Mobile users will have the ability to access the Internet and service providers by their proximity alone. Interfaces will extend beyond that of human-computer interface to include autonomous device-to-device communication. Merely walking by a particular cluster of appliances could allow your Bluetooth device to be probed for information that you may or may not want to disclose. Similarly, walking near a vendor or advertiser will open a connection to your device, which these parties could use to push desired or unsolicited information your way. You might say that the problem could be eliminated were the user required to enter a PIN to authorize the exchange. Yet this solution poses a challenge that compromises the system's core concept and functionality; it would be immensely irritating for a user to repeatedly have to enter the PIN, using one, or possibly multiple, Bluetooth-enabled devices. Further, the user might already be out of range before even getting the opportunity to enter the code.

Still, Bluetooth poses a considerable security risk because users can be completely blind to what is happening to their mobile device as they travel from place to place. Stealth viruses are poised to take advantage of this new medium and behave in the same manner as airborne viruses, jumping from device to device. In the past, users infected their computers with viruses by sharing information through the physical medium of a floppy disk or by downloading an infected file from the Internet. With Bluetooth, you just need to walk by someone. Picking up a mobile device virus from a stranger may become easier than contracting the common cold.

Although the industry is not fully developed to provide the utopian electronic community where consumers or business people roam freely and share data, users will quickly embrace the technology for its merit of freedom. The implications for m-Commerce transactions using this medium are staggering. Technologies such as Bluetooth are expected to significantly propel and enhance all forms of m-Commerce, especially in the consumer-to-consumer market. Accordingly, security risks will increase at the same rate and must be met with strict resolve, antivirus technology, and personal security policies.

In addition to catching a mobile virus, Bluetooth technology allows one to track users by their PIN numbers. Mapping a PIN to a user's identity opens the door for a whole new list of potential violations against personal freedom. Tracking the whereabouts of a user can provide a corporation with insightful information about consumer habits. Similarly, the technology opens channels for government to keep tabs on the citizenry. The threat of such information falling into the wrong hands and being misused increases many fold over previous threats associated with conventional

Internet access devices. Although security approaches are under development, manufacturers must focus on educating the users of their products in an effort to create awareness and to take precautions to protect their information and to keep it secret.

Infrared

Another interesting short-range wireless technology is known as infrared, or the IrDA (Infrared Data Association) standard. Infrared technology uses light waves to send and receive data. Most laptops, PDAs, printers, and a number of other mobile devices come packed with an infrared port ready for use. The trick with using infrared as a method of sending data is that you must have a line of sight of within a few feet for the signal to catch. It is however, extremely useful for sending data to printers when a USB or parallel cable is not readily available.

The functions of infrared are much more extensive than the authors of this book can possibly discuss in this text. Suffice it to say that infrared is currently used in photo and video cameras, defensive weapons, and a number of other applications that we won't go into. From a mobility perspective, infrared technology will always be an extension of communication and connectivity with other devices.

Radio Frequency Identification

Radio Frequency Identification (RFID) is most commonly used in RFID tags that store data about an object or a person. RFID technology falls into a category of mobile devices called "unintelligent gadgets," the thought being that RFID tags or devices contain little to no processing power. Their intent is to store data about a customer, a machine, another device, or any number of objects. At first sight, RFID sounds like it serves a very small purpose, when, in fact, this technology has tremendous uses in the mobile world. An example of RFID technology is the tollway system. Many of us that use the tollway system in our state to get to work every day understand that a price is charged for the use of the roads. RFID tags are used as small decals (sometimes small white boxes) that are placed inside the vehicle. When the car drives near the toll both, instead of stopping to pay the change required to pass, the RFID tag is read by a transmitter and opens the gate or simply allows you to pass in a gateless toll. The RFID tag stores the cost of passing through, which is sent to the patron in a monthly bill. Again, the RFID tag is a simple device with little intelligence, but in fact saves a great deal of time and speeds traffic significantly. RFID tags are also used in parking garages for patrons who park on a monthly basis and are required to pass through a privacy gate that would normally dispense a ticket. With an RFID tag, the customer information is stored on the device and read by the transmitter, allowing the vehicle to pass through into the parking lot. Other examples of RFID are water meters, which traditionally required access to someone's home to obtain readings on water usage, and other utilities. In this case, an RFID tag is placed in the proximity of the water meter, storing the data that is ready to be transferred when the time comes.

Finally, RFID technology stores critical marketable information about customers. RFID tags are being used all throughout the retail community to purchase gasoline, goods and services such as food and pharmaceuticals, as well as any number of other retail purchases. The RFID device essentially replaces the credit card used to make transactions. The convenience to the consumer is increased significantly. An additional business driver around RFIDs is the plethora of data captured about customers that would normally never be collected. Everything including age, sex, buying habits, and so forth can now be obtained to create additional marketing opportunities that never would have existed.

Satellite and Fixed Wireless

An alternative to DSL and cable broadband Internet connectivity is satellite and fixed wireless connectivity, typically marketed for consumer and small business use. Many of us live in areas where DSL or cable Internet is simply not available for one reason or another. DSL requires a location that is in close enough proximity to a telecom provider's base station. On the other hand, cable Internet networks are still in the process of expanding to markets throughout much of the North American continent. All is not lost however, because satellite and fixed wireless came to the rescue. The idea behind satellite communications is placing a small dish outside of the residence or building with a direct line of sight to a satellite network in the sky. To be more specific, the satellites reside several miles in space and typically require a direct line of sight to the south. Some cable satellite providers are offering Internet broadband services along with satellite television, which provides a very nice, package-based service. The speed of satellite transmission typically falls in the range of 256 to 800 Kbps download, and varying upload speeds, but usually around 56 Kbps to over 100 Kbps.

The other technology that is almost a hybrid of satellite technology is called fixed wireless. Again, fixed wireless is marketed as a high-speed alternative to wired connections that might not be available in some parts of the world. Fixed wireless works through a satellite-like device residing on the roof of a tall building or structure such as a high-rise building. The satellite communicates through a direct line of sight connection to as far as 45 to 50 miles in all directions to an endpoint connection. That connection is usually a small dish that is placed on the roof of a home or business and received the data.

The big difference in a fixed wireless scenario is that download speeds generally seem to be significantly higher. In fact, average download speeds can range from 500 Kbps to 7 Mbps! The large gap occurs because of the way that short bursts or packets data are sent over the air in a specific frequency. Each burst can be of varying speeds. At an average, however, the variances in download speeds are unnoticeable and work quite well as an alternative to wired connectivity. In fact, it is this author's humble opinion that fixed wireless can and does perform significantly faster than many wired connections.

Securing a Wireless Network

The security of the wireless network is no doubt an extremely important topic when considering whether mobile applications are right for your business. There are a great many factors to consider when transactions are sent over the airwaves as opposed to a wired connection. To provide secure transactions in any medium, there are four requirements that must be met: confidentiality, authentication, integrity, and nonrepudiation.

- **Confidentiality.** Requires that only the parties privy to the transaction be aware of that transaction's detail. In the mobile commerce environment, this suggests that the server and the terminal are the only two points that can examine the real contents of a transmission. To ensure confidentiality, the primary tool used is cryptography. Plaintext is encrypted at the origination point and decrypted upon receipt, effectively shielding the information from parties who do not have access to the encryption algorithm.
- **Authentication.** Asks that the parties in a transaction provide a means of proving their true identity. In the brick-and-mortar world, this is done through forms of trusted identification, that is, asking the trading partner for a prearranged "secret" code. In wireless data realms, the same concept applies. Server authentication provides a way for users to verify that they are really communicating with the entity with whom they believe they are connected. Client authentication verifies that the user is who he or she claims to be. Authentication can be implemented using passwords, tokens, and digital certificates.

- **Integrity.** Ensures the detection of any change in the contents of a transaction. In offline commerce, integrity has been accomplished by sealing documents and, in extreme cases, by providing a chain of custody. For the digital domain, analyzing transmission contents at reception and using algorithms that determine whether the content has been altered guarantee integrity. In addition, a digital signature can be used to provide a stronger test for integrity.
- **Nonrepudiation.** Demands that a party to a transaction cannot falsely claim that they did not participate in that transaction. In the traditional business arena, this is accomplished via signatures, seals, and notaries. In the wireless environment, nonrepudiation is more difficult to realize, although popular solutions include using a combination of digital signatures and certificates.

Wireless Equivalent Privacy

The 802.11 standard describes the communication that occurs in wireless local area networks (LANs). The Wired Equivalent Privacy (WEP) algorithm is intended to protect wireless communication from eavesdropping. A secondary function of WEP is to prevent unauthorized access to a wireless network; this function is not an explicit goal in the 802.11 standard, but it is frequently considered to be a feature of WEP.

WEP relies on a secret key that is shared between a mobile station and an access point/base station. The secret key is used to encrypt packets before they are transmitted, and an integrity check is used to ensure that packets are not modified in transit. The standard does not define how the shared key is established. In practice, most installations use a single key that is shared between all mobile stations and access points. More sophisticated key management techniques can be used to help defend from their decryption.

Summary

As we come to a close of our discussion of wireless networks, it's important to remember that we've only touched upon a topic area that contains volumes of information that is beyond the scope of this book. As wireless service providers, application developers, and integrators progress in the development and implementation of these technologies, the evolution process will be one to watch intently. Hopefully, this chapter has provided a good general understanding of where wireless networks came from and where they are going in the not-so-distant future.

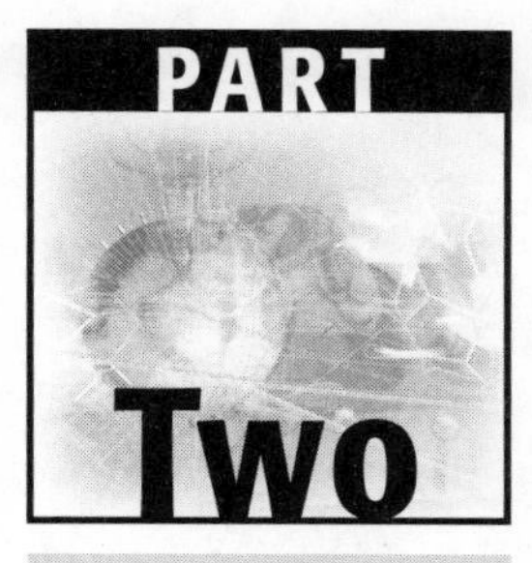

Opportunity Identification and Case Studies

CHAPTER

5

The Value Web Framework

Jorn Teutloff

Introduction

The previous chapters of this book provided you with comprehensive background information about wireless and mobility concepts. You learned about the unique attributes that distinguish wireless from mobile, read about various consumer and business drivers behind their rapid growth, and learned about some of the enabling technologies that make it all work. Based upon this foundation, we then provided you with a quick review of today's most prevalent mobile devices and the networks they operate with. For some of you, we hope that Part I of this book served as a useful refresher. Other readers may join us right now after having skipped through the background material. Welcome to Part II!

Defining the Value Web Framework

In this and the following chapters, we look at a unique approach to assessing the wireless universe. We will introduce Cap Gemini Ernst & Young's proprietary methodology for strategic analysis called the *Value Web*SM. The Value Web is an innovative framework intended to support CGE&Y project teams in the development and delivery of insights for the company's clients concerned with deriving value from mobile technologies. Value for CGE&Y's clients is usually defined as reducing operating expenses, generating additional revenue, and streamlining business processes to increase profitability while enhancing customer service. Thus, the framework is integral in the quest for understanding and applying mobile and/or wireless technologies for the purposes of creating a competitive advantage and improving the customer experience.

This chapter defines what we mean by a Value Web and presents the major benefits associated with preparing such an analysis as a tool for strategy formulation. We will then provide you with step-by-step instructions for how to create and analyze a Value Web for your own company. Chapter 6 presents an already completed Value Web for the wireless industry as we know it today. After you have read both chapters, you will see how the approach can provide valuable insights, whether it is applied at the company or at the industry level.

What Is a Value Web?

The Value Web is a framework, a structure, whose primary benefits include the channeling of thought during the process of strategy formulation. However, the concept represents more than just another variation of existing strategy tools that are commonly used in business today. The Value Web takes a completely different view of a company's business environment.

Traditionally, companies perceived themselves as a link in what is commonly known as a value chain. In the 1970s, the prevailing model was that of "one company does it all." Big organizations, such as IBM, DEC, or Wang, were vertically integrated, controlling the value chain all the way from sourcing and production to distribution and end-user support. Realizing the inefficiencies inherent in such a structure, industries gradually adopted horizontally integrated structures. During the 1980s and 1990s, companies moved from the "one company does it all" mindset to that of niche players, aligned along a horizontal dimension. To stay with the example of high-technology companies, Intel, Motorola, and Texas Instruments focused on designing the chips, Microsoft and Novell provided operating systems, IBM, HP, and Sun built computer systems, whereas Microsoft, Lotus, Oracle, SAP, and others developed the applications.

The Value Web framework is a fundamental departure from the traditional value chain concept, be it vertical or horizontal. Characterizing companies in the 2000s and beyond, the Value Web framework considers an industry or a market and its constituents to be represented best by the analogy of a network, or a web, rather than a linear sequence. In its essence, the Value Web framework describes a customer-focused and company-coordinated network that brings together strategic relationships that are required to provide the customer with a *total solution*. Companies that effectively coordinate their market's Value Web are the ones that will create a sustainable competitive advantage. Companies that are unable to design and coordinate an effective Value Web will be relegated to the status of low-margin commodity providers.

The rationale behind employing the analogy of a web, or a network, as a structure in strategy formulation lies in the fact that in today's complex business environment there are very few companies that can, on their own, deliver a complete solution to meet their customers' requirements. By a complete, or total, solution we mean an offering that consists of all products, services, and information that fulfill a customer's wants and needs. Instead of trying to go it alone, today's companies operate in a business environment where each constituent, each company, provides a unique value, all of which when taken together create the complete offering.

Let's explain using a simple example. When Henry Ford started to mass-manufacture the Model T in 1908, the output of his operation was a vehicle, a mode of transportation, developed and produced for the consumer. The relationship of the company with the consumer was strictly transactional in nature; Ford secured the raw materials, built the manufacturing plants, produced the automobile, and sold the product to a customer. The relationship between the company and the end-user was largely transactional in nature. The two parties to the transaction never crossed paths again unless a service issue arose or until the next purchase. Today's automobile purchasing process, however, is vastly different. What once was a simple, linear relationship between a company and its customers has evolved in a complex web of multiple parties interacting with each other in a continuous, nonlinear fashion. The parties to this process include several companies, many only marginally related to the physical product. To illustrate, some of the constituents of today's automobile Value Web include not only the car manufacturer, but also companies such as Edmunds that provide automobile reviews considered by the customer during the decision-making process, institutions such as GMAC that provide dealer financing, dealerships, insurance companies, service and repair providers, vendors of after-market accessories, and others. Anyone who has purchased a new car before can relate to this example, an example that doesn't even consider the multitude of networked companies on the supplier side that provide the inputs required to produce the actual automobile.

Why this extension, and as some might say complication, of what used to be a very simple process? The answer is customer demand. Today's customers base purchasing decisions on a variety of elements associated with the solution, including the physical product, associated services, and supporting information. Each of these factors is being considered throughout the customer experience life cycle, which ranges from prepurchase need identification to postpurchase support and satisfaction evaluation. Companies, understanding the complexities of the decision process strive to provide a comprehensive solution, yet at the same time realize that it would be impossible for them to provide each and every element of this solution in a profitable manner. Instead, all those various companies contributing the to automobile purchasing process, to stay with our example, form what can best be described as a cluster of providers, each aiming at the same customer.

As long as these contributing entities operate in isolation, we are indeed looking at merely a group or cluster of companies. Their only common ground is the customer whom they are pursuing for financial gain. To make the transition from a cluster to a true Value Web, these companies must become aware of each other and determine how to work together in providing the customer with a seamless experience. Working together, whether through formal or informal relationships, illustrates the central premise of the Value Web. Those companies that are able to coordinate and lead these webs of value providers are the ones that will capture a sustainable, competitive advantage. Companies that continue to operate in isolation are at risk of being pushed aside by those competitors who become part of the larger entity.

The Web's Anatomy

Realizing that the business environment is evolving from linear value chains toward interconnected Value Webs in response to customer demand, we might ask what is positioned at the hub of the construct. Whereas the more traditional approaches to strategy formulation place the organization at the center of the universe, while defining competitive advantage along the lines of internal capabilities and external opportunities, the Value Web starts with the customer. Therefore, the customer is located at the core of the construct—the center of the web—with every business decision evaluated against its impact on customer value. Competitive advantage is based upon a company being effective at coordinating partnerships and alliances that, in the aggregate, cater to the customer. The intent is to deliver the total solution by bringing together and orchestrating a supporting cast of companies that provide the functions that are not the core competency of the coordinating organization.

Moving outward from the center of the Web, we encounter the various constituents of the network. Located in the first orbit around the core is the company that provides the offering a customer is interested in, such as an automobile manufacturer. This is usually the company for which a Value Web analysis is performed; rarely do we see a client commission CGE&Y with the full analysis of another entity unless a merger or acquisition situation is being addressed. To distinguish this company from all the other entities in the web, we will refer to it as "the firm." Besides the firm, other Value Web constituents include the firm's direct and indirect competitors, suppliers, service providers, retailers and resellers, complementors, affinity groups, and independent authorities. Figure 5.1 illustrates the network.

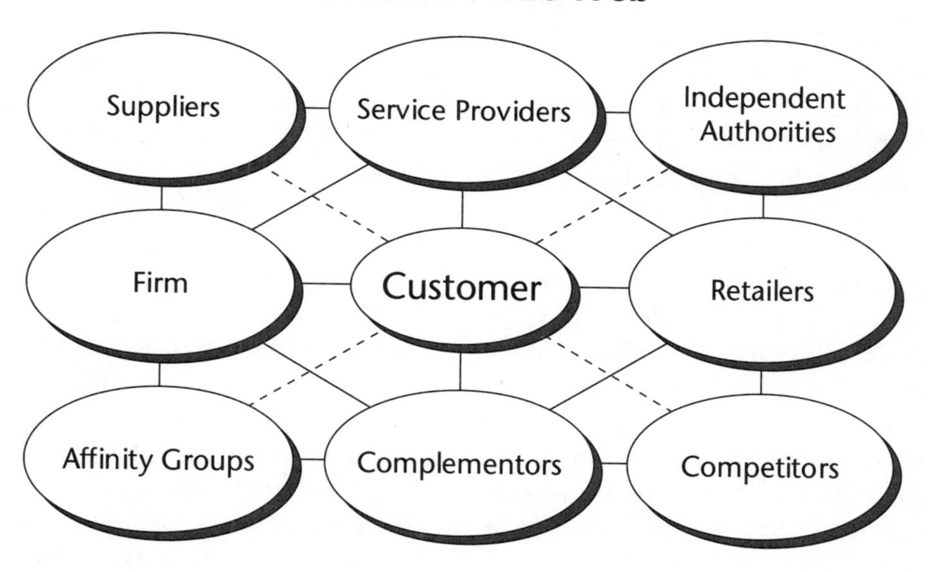

Source: CGE&Y Analysis

Figure 5.1 The anatomy of a generic Value Web.

Without spending too much time on the description of the concept, let's quickly define the constituents of the generic web presented in the figure. The firm, of course, is the provider of products and services targeting at fulfilling the customer's requirement. Competitors include those entities that provide a substitute offering, such as a product or service that provides the customer with the same or similar experience and utility as the firm's offering. Suppliers are vendors that provide the inputs that in their aggregate make up the firm's products/services. Service providers are those parties that fulfill critical activities that lie outside the firm's core competencies. Retailers and resellers act as the distributors of the firm's products/services to the customer, whereas complementors provide a separate offer that can be used in combination with the firm's products. Affinity groups—for example friends, family, and coworkers—influence the pre- and postpurchase process because they have some usually noneconomic link or affiliation with the customer. Finally, independent authorities, such as the media, governments, or other information providers, also participate in the Value Web even though they normally have no vested interest in the final transaction.

Figure 5.1 illustrates a rough, generic model of what a retail industry, customer-centric Value Web might look like. However, the Value Web framework lends itself to analyzing a given business environment from multiple angles. Instead of limiting ourselves to applying the framework to the macroeconomic level that considers entities external to the firm for which the analysis is performed, we could apply the structure to some of the firm's internal, operations-related processes.

A Value Web for a theme park operation, for example, would still place the customer experience at the center of the web. The constituents of the web, however, could be the following entities or activities: transportation to the location, overnight lodging, the admissions process, the park's rides and shows, merchandising, food and beverage services, guest services, off-site activities, and others. As you can see, the concept is the same, but this example is taking the approach to a different level. Both approaches have the same goals in mind: to understand the market space and to identify strategic opportunities to profitably create or enhance the customer experience.

Before we jump into the exercise of actually constructing a Value Web following a step-by-step approach, let's quickly review the reasons for doing so.

Why Use the Framework?

There are two major benefits associated with a Value Web analysis (see Figure 5.2). First, going through the exercise of developing and completing the Value Web framework for any given firm offers strategic insights into the firm's business environment. A completed Value Web allows its users to better comprehend the complexities of the conditions the firm is operating in by laying out the entities and the strengths of the connections between them. Secondly, identifying Value Web constituents and mapping their relationships is the foundation for formulating corporate and competitive strategies. Only when the network is understood can the firm begin to identify, pursue, and exploit strategic opportunities to enhance the customer experience while improving the firm's competitive position.

Value Web Applications

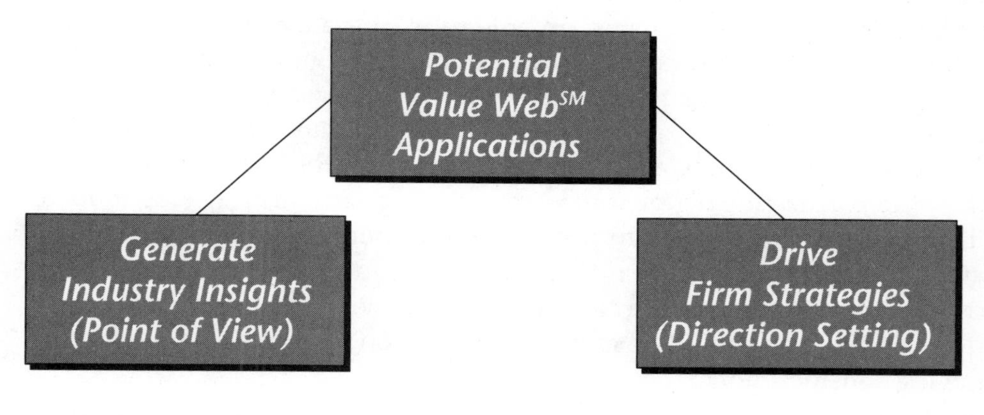

- **Helps develop detailed understanding of industry structure and "power" relationships between constituents**
- **Focuses strategic attention on major technological trends that may significantly affect industry structure and "power"**

- **Identifies key strategic issues for firm**
- **Analyzes firm's strategic "power" relative to competitors, and identifies key strategic initiatives**

Source: CGE&Y Analysis

Figure 5.2 Value Web applications.

On the external side, insights gained from analyzing the completed framework can be used to better understand the firm's general industry, a distinct market space, incumbent constituents, and their relationships. Among the constituents we count, as previously mentioned, competitors, suppliers, service providers, retailers and resellers, complementors, affinity groups, and independent authorities.

When applied to the firm's internal operations, a Value Web analysis can discover relations and interdependencies among functional departments, strategic business units, or multiple organizations within a multibusiness conglomerate. Company-internal constituents may include functional departments such as Human Resources, Sales, Marketing, Legal, Support, and Information Technology. Or they might entail operations segregated by geographic boundaries such as the Americas, Europe, or Asia-Pacific.

Whichever level the framework is applied at, the Value Web is an excellent tool to understand the landscape a firm is operating in by discovering and mapping "power relationships" among the web's key players. As the web emerges over time, users are able to map the construct's evolution, which is critical for strategic and operational realignment in response to changing market needs. Thus, the Value Web provides a comprehensive picture of past, current, and potential future corporate interdependencies.

In addition to gaining a better understanding of the business landscape, being able to see the big picture is important for any firm seeking to identify opportunities to satisfy unmet customer demands. After having created a Value Web, a firm is better positioned to formulate its corporate and competitive strategy. At the corporate strategy level, insights gained from drawing the map provide the firm with decision inputs relating to the following considerations:

- Which markets to target
- Which acquisition or divestiture opportunities to pursue
- How to allocate resources among competing business ventures
- Which entities to form strategic alliances or partnerships with
- Whether or not to vertically integrate, and so forth

At the level of competitive strategy, the Value Web allows you to better determine how to compete. Establishing a sustainable competitive advantage, which is the ability to create profits that exceed those of the competition, can be accomplished by identifying opportunities for superior cost management or product/service differentiation. The Value Web assists with identifying opportunities to outsource noncore activities to its network of relationships, while focusing on core competencies. It helps the firm to assess its strategic position versus other players, and to determine prioritized, strategic initiatives with the highest potential for significant returns.

Last, besides allowing a firm to survey its environment, the constituents, their relations, and the associated strategic opportunities, the Value Web also focuses attention on major trends that may affect the equilibrium. The disturbances are often of competitive nature as in the case of new market entrants, but they could also be of technological nature, stem from government-initiated deregulation efforts or shifts in cultural trends, and so forth. By grouping the Value Web's members according to the service they provide and by arranging these groups in such a way that interdependencies become visible, a firm can attempt to discern the effects of new breakthroughs upon the balance that holds the web in place. Again, leveraging insights into potential or acute disruptions of the status quo will provide the Value Web user with the ability to change course accordingly.

As we said before, the goal for the firm is to coordinate the web, to manage and lead it, by orchestrating the constituents in an effort to enhance customer value and create competitive advantage. The Value Web analysis provides an effective foundation for any strategic project. Once the web is completed, strategic insights have been created, and a set of potential projects identified, subsequent activities include the development of business cases to evaluate each project, and the planning of an implementation process, as shown in Figure 5.3.

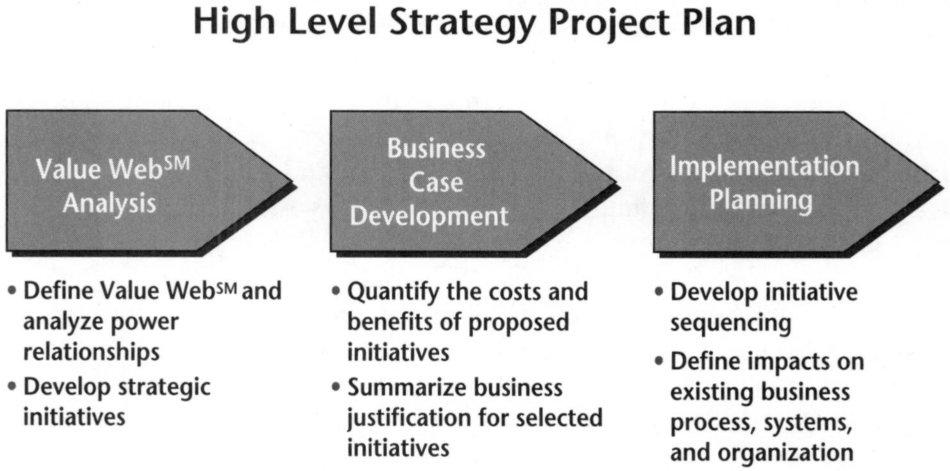

Figure 5.3 High-level strategic project plan.

The following sections will focus on creating and analyzing a Value Web for a fictitious firm. The steps we will outline are the same for any organization.

Constructing a Value Web

Now that we have defined the Value Web and described the benefits for strategy formulation, let's look at how a firm would actually go about constructing the framework. In the next sections of this chapter, we will walk you through the steps required to develop and complete a Value Web. We will follow the same approach that our consulting teams use when creating the analysis for one of CGE&Y's clients. A little later, we will explore how to analyze the completed framework to distill insights that we then can use to set a firm's strategy.

Before we jump into the details, let us quickly review some of the guiding principles to be applied during the exercise. First, as mentioned earlier, the focal point of the Value Web is the customer. Located at the center of the framework, the customer is the ultimate recipient of value; only if there are unmet customer needs or wants will companies strive to provide an offering that will fill the void. Thus, all activities in the web should be focused on delivering value to the customer.

Second, the exchange process forms the basis for creating relationships among the web's constituents. When we talk about providing "value" to the customer, we refer to an offering that may consist of a physical product, a service, information or any combination thereof. Providing such an offering takes the form of an exchange, where a firm gives something of value to a customer, usually in return for a monetary payment. Exchanges, however, also take place between web constituents other than the customer. Those exchanges, for example between the firm and an independent service provider, usually entail information. Thus, exchange processes illustrate the relationships, or linkages, between Value Web participants.

Third, the scope of an initial Value Web analysis should be bounded by first order relationships. By this we mean that our focus is to identify the critical constituents and to map obvious power relationships to be used as the basis for further analysis. Second and third order relationships are analyzed as required; for example, when the firm is interested in discovering constituent-specific strategies, possibly in the form of a detailed competitor analysis.

Let's now look at the individual steps for creating the web. To render this chapter less of a strictly academic model, and provide you with a more practical application of the Value Web concept, we will walk through the steps from the perspective of a fictional company called Acme. Acme, a no-frills provider of diversified financial products and services, wants to use a Value Web analysis for the purposes of identifying opportunities to use mobile technologies. The steps that Acme follows in creating a Value Web include:

1. Defining the solution and competitive space
2. Selecting target markets and market segments
3. Defining customer needs and identifying customer experience life-cycle constituents
4. Defining value transactions

Step #1: Defining the Solution and Competitive Space

The first step in creating a Value Web is to identify the firm's offering, or unique solution, in relation to the total solution. If you remember, a total solution, as defined earlier, provides a customer with a complete set of required or desired functionality, including pre- and after-purchase needs. We will talk more about the total solution in step #3, using the customer experience life cycle. For now, the firm should focus on identifying and assessing all the value propositions that it can realistically offer to its customers via the products or services it provides. The firm's offering, its unique solution, then, is defined by the utility provided to customers; why, how, when, and where this offering is being used. Additional questions that can be useful when defining the unique solution include those that investigate the acquisition process: Where, when, and how are customers buying the firm's products? The result of this exercise is a good description of what the firm goes to market with.

In addition to clearly defining its offering, a firm needs to thoroughly assess its competitive space. At the highest levels, the firm needs to identify providers of substitutes—products or services that fulfill the same or similar needs as the firm's offering. Questions that help in this exercise include: What other companies offer the same or similar products? How are these products different? How are they competing with us in terms of internal business processes? How are these competitors going to market? What are their short-term and long-term strategies? What other companies could infringe upon Acme's turf by adjusting their product mix? What are the chances of new entrants setting up shop? Answers to these types of questions need to be collected and thoroughly analyzed in a fashion similar to a thorough, traditional competitor analysis.

Let's look at these steps using our example. As we said earlier, Acme is a diversified financial services firm, offering no-frills brokerage, banking and insurance products, predominantly using an Internet-based self-service distribution model. To keep this example manageable, we will focus on only one of Acme's product lines, the firm's online brokerage business. For our purposes, then, the unique solution (online brokerage) Acme provides to its customers consists of the firm's marketing activities that create consumer awareness, its network of financial advisors, the Internet Web site through which most of the firm's business is conducted, and its few branch offices and call centers that also interact with the customer. In addition to these externally focused services, Acme's offering also entails the actual trading functionality that allows a customer to buy or sell securities. Acme's competitive space is characterized by other financial institutions offering competing products and services. Such competitors include small, no-frills startups all the way to the traditional full-service brokers. In addition, there are nonfinancial substitute products that could serve as investment vehicles, including high-end art, real estate, and other tangibles that are expected to appreciate over time.

Step #2: Identifying and Evaluating Market Segments

Once Acme has defined its offering and competitive space, it must segment the general market and group like segments to arrive at solution areas. To illustrate, lets continue with our example. We said we would focus on Acme's online brokerage as the unique solution for our example. This solution can serve multiple customer segments. At the highest level, two broad customer segments may be businesses and end-consumers. If desired, these two segments could be further broken out in an effort to arrive at more detailed clusters. For our example, however, we will only concern ourselves with the high level business and consumer segments. Both segments, labeled A and B, form the solution area under the brokerage offering. Figure 5.4 illustrates this solution area in the first column of the left table. The second and third columns show two other solution areas, solution areas #2 and #3. The former consists of geographic market segments, possibly for Acme's banking offering, whereas the latter entails demographic segments, perhaps for Acme's insurance products. Note that these two solution areas are included for illustrative purposes only—we will not refer to them going forward.

Having identified our targeted market segments, we now map them in a screening matrix. The screening matrix plots the potential value of a segment against the ease of market entry. Potential segment value is defined by the segment's size, growth, and profitability. Ease of market entry reflects Michael Porter's forces of industry competition, including competitiveness from incumbents and substitutes, capital requirements, access to distribution channels, supplier and buyer bargaining power. Figure 5.4 shows how the segments within the first solution area map using the screening matrix. According to this analysis, the Consumer segment (B) ranks higher on both potential value and ease of market entry, and thus should be the segment that Acme pursues first.

Defining Acme's unique solution and evaluating targeted market segments within solution areas will serve as the backdrop for the next step; identifying customer needs along the entire customer experience life cycle.

Step #3: Mapping Customer Needs and Value Web Constituents Along the Customer Experience Life Cycle

After defining Acme's offering and identifying the targeted customer segments, the firm is ready to define each segment's needs along the customer experience life cycle. This exercise is a critical step in defining what we called the *total solution*: the aggregate of all products, services, and information that cater to a customer's requirements. In addition, this step structures the firm's efforts to identify other constituents, other companies, who seek to fulfill the customer's unmet needs.

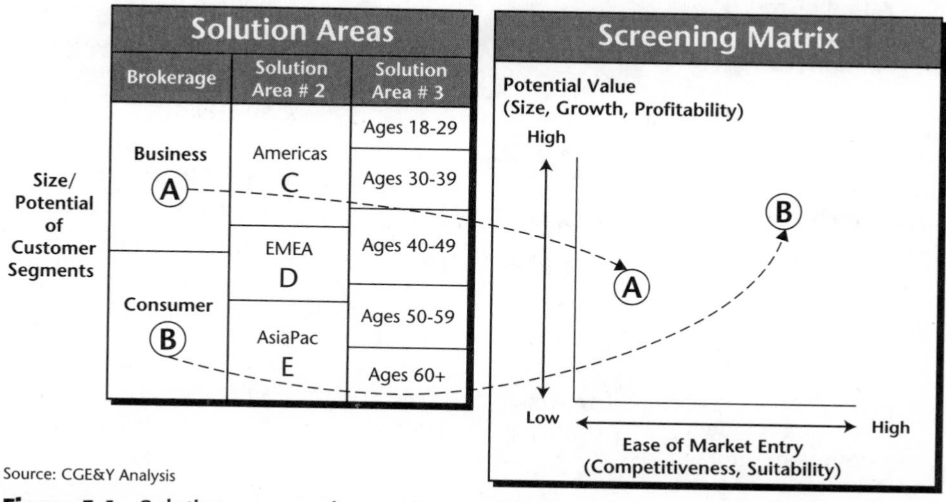

Figure 5.4 Solution areas and screening matrix.

There are six life-cycle stages through which a customer progresses when attempting to satisfy a need: need identification, research, purchase, receipt, support, and evaluation. Figure 5.5 illustrates the model. Note that we have listed the generic types of constituents most likely to be encountered in each life-cycle stage. These constituents include the firm that offers the product, competitors providing substitutes, service providers enabling a transaction, retailers and resellers distributing the product, complementors providing an offer to be used in conjunction with the product, affinity groups such as friends and family, and independent authorities as information providers. A firm undertaking the analysis will want to list the actual companies that fall within each constituency group.

The purpose of applying a life-cycle model is to identify the customer's requirements from initial need recognition to after-sale satisfaction assessment. Identifying the customer's needs and the specific constituents at each stage of the life cycle allows the firm to develop strategic initiatives that affect the experience life cycle. This is where the notion of a total solution comes up again. The life cycle helps us structure our inquisition into where, when, and how Acme's customers are becoming aware of the firm's products; when and how customers are acquiring these products; and how they are using them. Let us look at the individual life-cycle stages and the constituents that influence each.

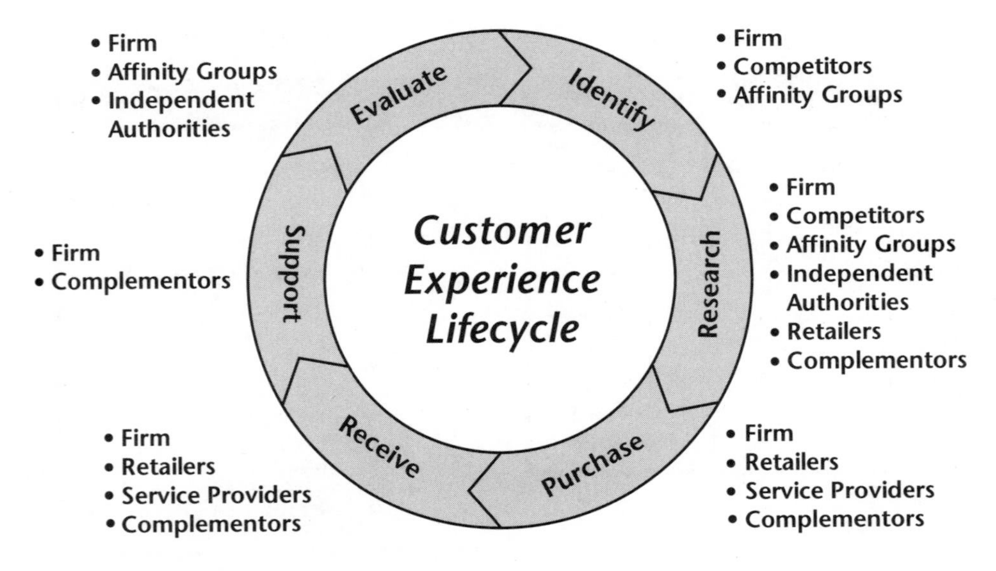

Figure 5.5 The customer experience life-cycle model.

Identify

The first stage of the life cycle describes customers recognizing an unmet need (or want), and their becoming aware of and identifying potential solutions. The constituents present at this stage usually are the firm, its competitors, and affinity groups.

To resume with our example, a consumer—let's call him Mr. Smith—realizes a desire to ensure his family's financial well-being over the coming years. Mr. Smith has lately become aware of this need, notably after the birth of his daughter, whom Mr. Smith and his wife would like to attend college and possibly graduate school. In addition, Mr. Smith is thinking about his retirement. Although it is a few years off, he would like to start planning for this event now to ensure the family's income once he stops being employed. Mr. Smith has started thinking about this ever since watching Acme's television commercials and reading through Acme's competitors' advertisements in the press. Last, but not least, his wife has recently begun to bring up the topic of financial planning for the future, and Mr. Smith now realizes that he had better look into this.

Research

Once a customer identifies a need, and the options to meet that need are clear, he or she evaluates them in the second stage of the model. Unless we are dealing with a low-value or repeat purchase situation, this process usually involves researching the options' value propositions, taking into account not only factors such tangible benefits and their price, but also intangibles such as brand image, accessibility, after-purchase support, and so forth. Affecting the customer's research process are various members of the Value Web. It is at this crucial stage that the customer comes to a decision that will lead to a purchase, which explains the presence of so many constituents. These entities usually include the firm (of course), competitors, complementors, and retailers whose goal is to participate in the financial transaction. Other constituents, usually lacking the financial profit motive, include affinity groups and independent authorities such as the media.

Continuing with our example, Mr. Smith has set out to learn more about financial planning. He decides to research companies that offer financial services that would allow him to build a college fund for his daughter, buy life insurance, and build a portfolio of securities. As part of his quest for information, Mr. Smith searches the Internet and finds several financial Web sites that offer advice on all three topics. During this search on building a stock portfolio, he comes across Acme Financial as a brokerage that offers a no-frills, low-cost, securities trading over the Internet; this is very much to Mr. Smith's liking. Mr. Smith scans Acme's Web site, and orders additional brochures from Acme's Marketing department. During his Web search, Mr. Smith also comes across Acme's brokerage competitors, all of which are similarly eager to provide Mr. Smith with the information he seeks and secure him as a customer. In addition, Mr. Smith learns about companies that provide complementary services that would allow him to better manage his portfolio were he to build one, and he finds independent organizations that operate Web sites that rate the various brokerage houses and their financial stability. Last, but not least, Mr. Smith turns to affinity groups, including his friends, family, and colleagues, to obtain their opinions about the benefits of investing in the stock market, what online brokerages they are using, and how satisfied they are with their brokerages' service offerings and customer support.

Purchase

The result of the customer's evaluation of options is a prioritized list of alternatives, the highest of which the customer selects and purchases. Ideally, the entities involved with the third stage of the life cycle, the actual purchase process, are the firm from which the customer buys, service providers that enable the transaction, complementors, and the retailer if such an intermediary is required to transfer ownership of the asset. Indeed, with certain product categories, including consumer electronics, for example, the purchase process is handled entirely via the retailer and not the manufacturing firm, which explains the firm's potentially being absent in this life-cycle stage.

Referring back to our example, Mr. Smith has arrived at the decision that Acme Financial is the brokerage that offers him the most value in terms of product range, price, and customer services. He opens an account with Acme, and transfers an initial sum from his bank's checking account. At the same time, Mr. Smith purchases a software application for his home computer that offers him comprehensive portfolio tracking to enhance Acme's bare-bones securities trading functionality.

Receive

The fourth stage in the customer experience life cycle pertains to the receipt of the offering, which might not include its installation by the provider.

Again, those parties that were privy to the transaction—the firm, complementors, service providers, and retailers—are usually involved in the delivery of those products and services, although this is not a requirement. Frequently, firms completely outsource the delivery function to resellers and retailers, and thus are absent in this stage.

For Mr. Smith, who is now a registered customer of Acme's online brokerage, receipt of the product or service entails having access to Acme's online self-service trading desk, as well as the firm's branch offices and brokers who Mr. Smith can contact via Acme's toll-free telephone number.

Support

The consumption of the offering is supported by after-sales service, usually provided by the firm and complementors in the fifth stage of the customer experience life cycle. For Mr. Smith, brokerage support entails reading the Acme Web site's detailed instructions about how to perform a trade, viewing an animated online demonstration, perusing the site's frequently asked questions section, or calling the firm's 24/7 customer support hotline.

Evaluate

The sixth and final stage of the life cycle is the evaluation stage, during which the customer's satisfaction with the offering is assessed. Mr. Smith is very happy with Acme, and tells his friends at work about his positive experience. He also responds to the online survey Acme sends him from time to time, inquiring about his experience and how to make the site even better. Last, but certainly not least, Mr. Smith is proud to present to his wife with a solution to the need for financial security that—in combination with a college fund and life insurance to be secured next—will further stabilize the family's financial future.

Step #4: Defining Value Transactions

Having identified all relevant life-cycle stages, their constituents, and the value each brings to the customer, we are now ready to draft Acme's Value Web, which will illustrate the types of transactions our constituents engage in. Figure 5.6 shows a completed Value Web for Acme's brokerage offering.

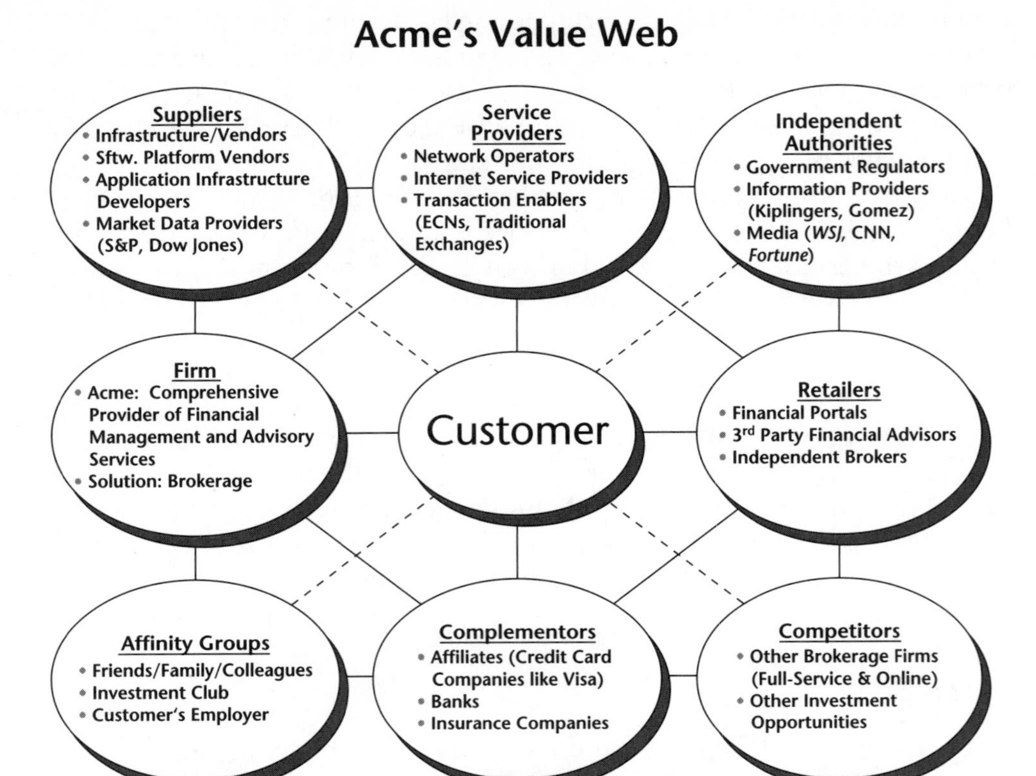

Figure 5.6 Value Web for Acme's brokerage.

Note that we are using solid lines between constituents when they are interacting at a close level, usually in the form of an exchange of a tangible offering. We distinguish these *transactors* from *influencers*, represented at the nodes of a dotted line. Dotted lines denote indirect or nonfinancial relationships that are of an influential, rather than a physical, nature. For example, although suppliers interact directly with the firm by providing raw materials required for production, they rarely engage in customer-facing activities. A customer consulting with independent authorities is captured via a dotted line because no financial transaction takes place. Similarly, the customer's relation to the firm's competition is indirect, because by definition for our Value Web analysis the competitor is not the entity that the customer will end up purchasing from.

Now that we have completed the customer life cycle, identified the customer's needs at each stage, listed the companies that seek to address each need, and defined the constituents' value transactions, we can begin to analyze the work completed thus far and draw some conclusions.

Leveraging the Value Web

The whole purpose of constructing the network is to look for insights that will allow us to understand the power relationships among Value Web constituents and to develop a portfolio of strategic initiatives that provide us with a competitive advantage. The steps we will follow in leveraging our Value Web include:

1. Analyzing value and sustainability to define relative power
2. Defining strategic focus based on power shifts
3. Developing strategic initiatives and quantifying impacts

Step #1: Analyzing Value and Sustainability to Define Relative Power

To identify potential strategic actions relative to its business environment and current or future players, the firm conducting the analysis should classify the relative power of the web's constituents into what we call a *power grid*. The power grid illustrates constituent influence by placing the network members into a matrix showing the value and sustainability that each brings to the web. The intent of the power grid is to quantify the relationships among Value Web constituents. The following sections illustrate how a power grid is created. The three activities involved include creating a value index, assessing the sustainability index, and creating the power grid.

Creating a Value Index

Our first activity is to construct a *value index*. The value index is a grid that quantifies the value each of the web's constituents receives and provides. We start by drafting a value transaction matrix that lists each constituent along the vertical and horizontal axes. In our example, we labeled the column header as "Outbound: Value Delivered to Others," whereas the row header specifies "Inbound: Value Received from Others." We then populated each cell in our matrix with the types, or dimensions, of value transactions between two constituents. These dimensions include Economic value (E), Informational value (I) and Intangible value (T). Economic value describes the exchange of goods or services usually for a monetary payment. Examples include raw materials, finished goods, or maintenance support. Informational value includes the exchange of content, customer data, or intellectual capital across nodes. Intangibles include influence relationships, opinions, emotions or other unquantifiable transactions between Value Web constituents. As seen earlier, an example of intangible value includes an affinity group or independent authority issuing an opinion that influences the buyer's decision-making process.

Figure 5.7 shows a partially completed, generic value transaction matrix. Note that some cells contain all three value dimensions—that is the two constituents transact with each other along the lines of a monetary, informational and intangible value—whereas other cells may show only one or two value dimensions. For example, the value provided by a firm to its customers usually involves all three value dimensions (E, I, T), whereas a customer interacting with affinity groups, such as friends and family, is mostly limited to the exchange of information and opinions (I, T).

Next, we determine the importance of each value transaction to the receiver. We assign to each value dimension a rating of high (5 points), medium (3 points), low (1 point), or not applicable. A constituent who provides value transactions that have a high importance to the receiver will have more power in the web than those constituents who are involved with low-importance value transactions. High-importance value transactions are those that have a significant impact on the solution, usually involving a high degree of customer-facing elements, whereas transactions of medium importance affect the overall customer solution less directly, perhaps due to the availability of multiple alternatives. Value transactions of low importance have little effect on the solution, and are characterized by little customer willingness to pay for or recognize them. Assigning these ratings to each value transaction is both an art and a science. If one is available, a firm may assign this task to their internal strategy group. More frequently, however, we see firms retaining an outside consultancy such as CGE&Y to prepare an objective assessment of the organization's value transactions.

Generic Value Transaction Matrix

E = Economic Flow
I = Information Flow
T = Intangibles Flow

Inbound: Value Delivered from Others	Outbound: Value Delivered to Others								
	Customer	Firm	Service Providers	Retailer/ Reseller	Comple- mentors	Affinity Groups	Suppliers	Independent Authorities	Competition
Customer		E, I, T							
Firm	E, I								
Service Providers	E, I								
Retailer/ Reseller	E, I								
Comple- mentors	E, I								
Affinity Groups	I, T								
Suppliers	I								
Independent Authorities	I, T								
Competition	E, I								

Source: CGE&Y Analysis

Figure 5.7 A generic value transaction matrix.

In addition to rating the level of importance of each value transaction, we need to determine the quantity of value transactions and include that measure in our matrix. In other words, we want our model to reflect the concentration of players in our web, realizing that the number of providers—and accordingly the number of value transactions provided—affects the level of value received. The more providers in a given constituent group, the more value the receiving party can expect to gain. Accordingly, our scoring index assigns a high (5-point) rating when there are many members in a constituent group. A medium (3-point) rating indicates a moderate number of value providers in the group, whereas a low (1-point) rating reflects a small number of constituent members. Again, it is crucial that ratings are based on objective, external research and expert judgment if the analysis is to provide valid results.

Now that we have determined the importance of each value transaction to the receiver, and specified the number of value providers, we can calculate the incremental value for each constituent group. Let us revisit the previous example. We will work with two cells: the one that contains the value a customer provides to the firm (E, I) and the one that contains the value he receives from the firm (E, I, T). Figure 5.8 shows how we calculate the total value for each of the two cells. First, we multiply the importance rating (importance of transaction to receiver) by the degree of concentration (number of providers) to arrive at the incremental value for each of the three value dimensions (E, I, T). Next, we add these values to arrive at the total value received. Then, we transfer this number to the corresponding cell in the value index table. Adding the values of each cell in a row (and column) yields the total value each constituent group receives (or provides).

The final task is to create a normalized metric that will later allow us to plot value index measures against the sustainability index that we will develop next. Dividing each row total by the corresponding column total produces the value index for each constituent group. Referring to Figure 5.8, the value index for the customer constituency equals the sum of the cell values for the row (15+X) divided by the sum of the cell values in the corresponding column (50+Y).

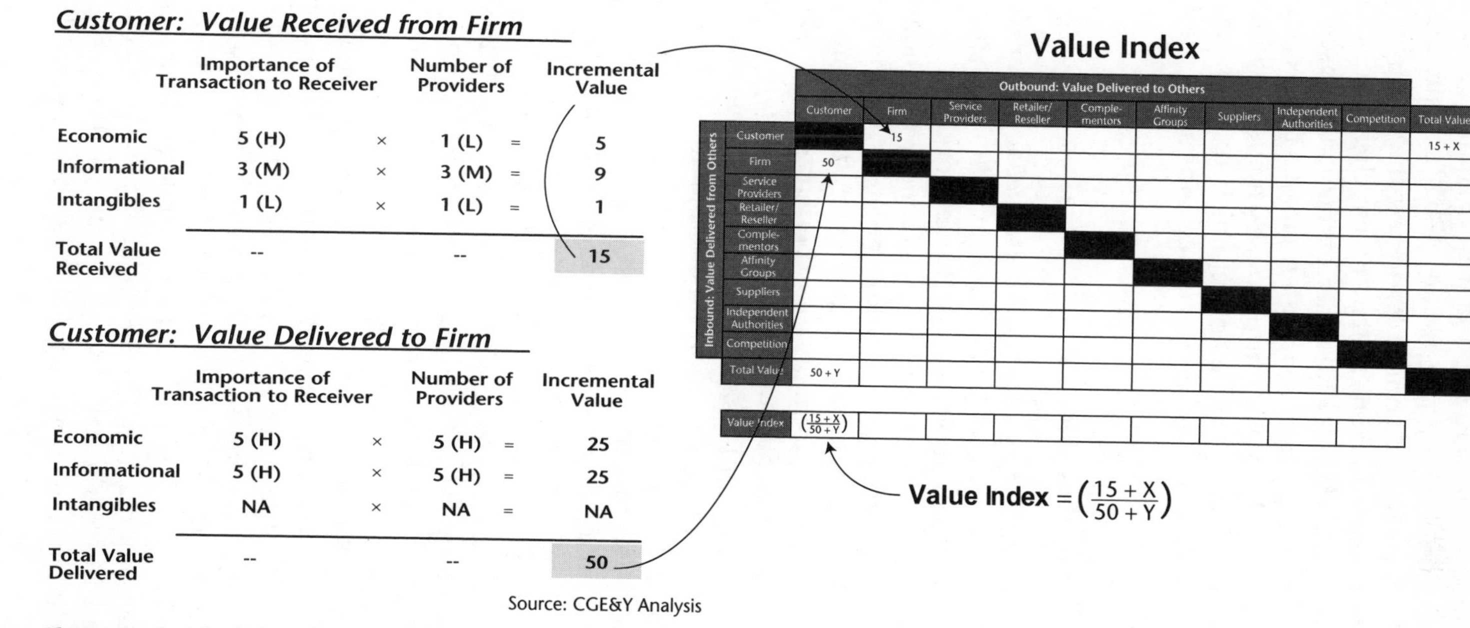

Source: CGE&Y Analysis

Figure 5.8 Calculating the value index.

Preparing the value index was important to quantify the nature and strength of the relations among the Value Web's constituents. Going through the effort of constructing and interpreting the value index alone can yield valuable insights into the extent of which network members receive (or provide) value from (or to) other participants. Based on these insights, a firm can chart in particular short-term tactics that exploit gaps in the existing value relationships. For our purposes, however, which was to assess the complete Value Web, we are not yet finished. The next step provides for the dimensions of constituent relevancy and ease of market entry to enter our analysis.

Assessing the Sustainability Index

Our next activity is to prepare and assess the *sustainability index*. This index tells us the staying power of each constituent. It contains two important measures: entry barriers and importance to the network.

When assessing entry barriers, we are concerned with understanding how susceptible a constituent group is to new members, mostly in the form of competitors. Characteristics to be considered include first mover advantages to favorable locations, technology, experience curve advantages, brand reputation, and equity. Startup costs, capital requirements, customer switching costs, access to suppliers, and distribution channels must also be considered. Additional factors include the adaptability of existing constituents to changing market conditions, as well as exit barriers. The scoring of entry barriers is straightforward: high (5 points) when significant barriers exist, thus making market entry unlikely; medium (3 points) when a reasonable, yet not insurmountable investment is required to become an effective player; and low (1 point) when few barriers to entry exist. As mentioned, defining barriers and assigning ratings is a delicate task that should be performed by experienced personnel and/or with the help of outside analysts to improve the reliability of the work.

The second measure is the constituent's level of network significance. Characteristics we want to investigate include the number and duration of direct and indirect ties with other constituents and the ability to forge successful alliances with other network members. A high (5 points) rating indicates the constituent's criticality to the network. A medium (3 points) rating shows that the member is important, but could be replaced by another player, whereas a low (1 point) score denotes those constituents that provide little unique value. The sustainability index is calculated by dividing each constituent's sustainability measure by the average sustainability of all constituents combined. Figure 5.9 shows a completed generic sustainability index, used for illustrative purposes only.

Sustainability Index

Constituent	Entry Barriers		Importance to the Network		Constituent Sustainability	Sustainability Index
Customer	1 (L)	×	5 (H)	=	5	0.6
Firm	3 (M)	×	3 (M)	=	9	1.0
Service Providers	1 (L)	×	1 (L)	=	1	0.1
Retailer/Reseller	1 (L)	×	3 (M)	=	3	0.3
Complementors	5 (H)	×	5 (H)	=	25	2.8
Affinity Groups	5 (H)	×	1 (L)	=	5	0.6
Suppliers	5 (H)	×	5 (H)	=	25	2.8
Independent Entities	5 (H)	×	1 (L)	=	5	0.6
Competition	3 (M)	×	1 (L)	=	3	0.3
Average Sustainability	--		--		9.0	--

$$\text{Sustainability Index} = \left(\frac{\text{Constituent Sustainability}}{\text{Average Sustainability}}\right)$$

Source: CGE&Y Analysis

Figure 5.9 A generic sustainability index.

Similar to an analysis of the value index, creating and interpreting the sustainability index on a stand-alone basis can reveal strategic insights. Yet, it is the combination of both the value and sustainability indexes that provides us with the biggest return on investment of time and effort thus far: the *power grid.*

Putting It All Together—The Power Grid

Now that we have prepared both the value index and the sustainability index, we are ready to take the last activity required to complete the power grid. Combining value with sustainability helps us define the power of the incumbents, anticipate power shifts, and develop effective strategies based upon these insights.

We will start by creating the basic grid. The power grid plots the value index as its Y-axis, while placing the sustainability index along the horizontal (X) axis. Figure 5.10 shows the power grid we developed for Acme Financial, the brokerage firm introduced earlier.

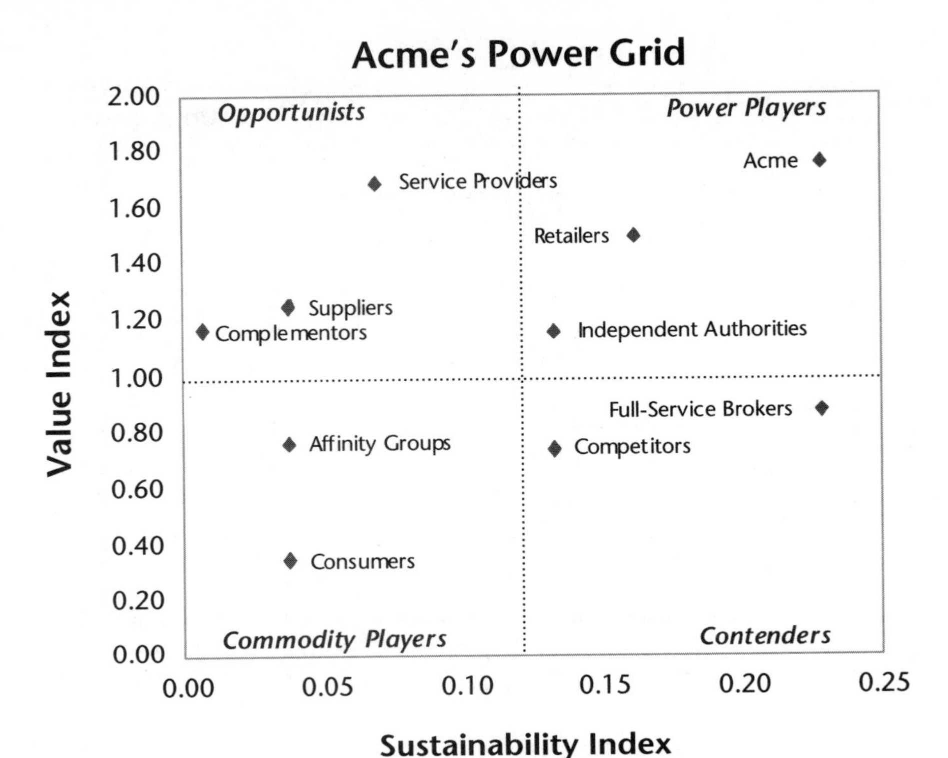

Figure 5.10 Acme Financial's power grid.

As you see, the power grid is a 2 × 2 matrix, consisting of four quadrants, labeled *Opportunists* (high value, low sustainability), *Commodity Players* (low value, low sustainability), *Contenders* (high sustainability, low value), and *Power Players* (high value, high sustainability). In the next step, we evaluate the position of the constituents and then start thinking about how to improve or fortify each position.

Step #2: Defining Strategic Focus Based on Power Shifts

Step #1 entailed analyzing value and sustainability. Starting with a completed power grid, step #2 will help us to develop the strategic direction for our firm, Acme Financial, and to anticipate the strategic moves of other constituents in the grid.

Opportunists, plotted in the power grid's upper-left quadrant in Figure 5.10, are providing critical value to the overall solution, yet their position is weak in nature and constantly threatened by newcomers to the game. In our example, we see service providers as the constituent group that finds itself in the opportunist quadrant. Service providers in our brokerage example are companies that enable or facilitate the transaction, for example by making available the network infrastructure required to buy or sell securities and settle accounts. Although providing critical add-on functionality to the overall solution, that functionality is usually not unique or sophisticated enough to keep at bay imitators that provide substitute offerings. The opportunist is successful at identifying and providing a critical element to the total solution, but unless he develops a strategy to distinguish himself from other members of his constituent group, his position is feeble, requiring the company to continuously refresh its offerings in a never-ending chase for survival. Thus, the appropriate strategy for the opportunist is to move from the upper-left to the upper-right quadrant. A set of strategic initiatives would primarily aim at identifying how the opportunist can attain a stronger foothold in the market via differentiation.

Low value and low staying power describes the *commodity players*, located in the lower-left corner of the power grid. Clearly situated in the least desirable position from the perspective of a company aiming to lead its market's Value Web, companies with commodity player status must make every effort to shift out of their low-margin corner or risk being pushed out of the market altogether. If they are providing a product or service, their contribution to the solution is not valued by the Web, nor is their staying power significant enough to allow them continued viability. Constituents located in this quadrant are at great risk of being displaced by stronger members of the network. Strategic actions include both a reevaluation of the company's product or service offering in an effort to increase their value-add, and deliberate tactics to increase the company's staying power (such as those attained through the creation of meaningful strategic alliances). In our example, we observe that both the consumer constituency and affinity groups are located in this quadrant, indicating their low power position in the brokerage industry. Both consumers and affinity groups are not strong enough to control the environment; their bargaining power is limited, forcing them to passively accept the offerings presented as opposed to actively shaping them.

Contenders, positioned in the lower-right quadrant, are enjoying a strong foothold in the Web, yet the value they bring to the solution is small and not viewed as critical by others. Our example shows the full-service broker, a specific member of the competitor constituency, barely contained in this quadrant. Acme's Value Web illustrates that the full-service offering, and associated price points, are valued less than Acme's no-frills offerings, yet the competitor enjoys the same strong foothold that Acme currently holds, perhaps due to its targeting a different market segment. Contenders should focus on raising the level of value their offering brings to the web, possibly by offering a web-based self-service model, a course of action many traditional financial services providers have investigated and implemented in an effort to keep pace with the Internet-only brokers.

Finally, the strongest position in our power grid is held by what we call the *power players*, located in the grid's upper-right quadrant. Power players are providing transactions that are viewed as valuable, while their position is strong and difficult to assail. Obviously, it is this position—and the underlying factors that provided a company with same—that allows their holders to attempt to take a leadership role in the organization of the Value Web. If you recall, at the beginning of this chapter, we stated that the strategic goal of a firm is to gain sustainable competitive advantage by coordinating the Value Web from near the center—versus simply being a member located on its fringes. Our analysis shows that the power players are in a prime position to attain that status.

Of course, there may be more than one constituent located in this quadrant, which would require the firm that undertakes the Value Web analysis to further refine its strategic approach to outmaneuvering other companies vying for the same leadership position. Such refinement may be attained through scenario planning exercises that identify future alternatives and would drive the creation of contingency plans. In our example, Acme currently enjoys a power player position by offering the highest value to its targeted customer segments—individuals who prefer online to broker-assisted trading.

Having assessed our own strategic focus and that of the other members of the Value Web, we can now take our analysis to the next and final level—which is to develop a portfolio of specific strategic actions.

Step #3: Developing Strategic Initiatives and Quantifying Impacts

The last step in our exercise is to develop a set of strategic initiatives. Because such initiatives are unique to any given firm and the industry it is operating in, we cannot attempt to lay out specific recommendations for your organization without further due diligence. However, we can present you with a set of generic actions that any company could undertake given its position in the Value Web relative to the other network participants. We hope that you will find this list of actions to be a good starting point for your own analysis and development of a custom-tailored initiative portfolio that is right for your company.

When facing a strong and powerful *customer* constituency, a firm's strategic actions may include efforts to continually differentiate the offering to bolster the firm's position as a high-value-adding provider. In addition to product differentiation, the firm may investigate the feasibility of partnering with or acquiring competitors to consolidate power, and/or create switching costs to lock in the powerful constituency. If the customer constituency holds a relatively weak position in the power grid, a firm's strategic actions could focus on raising prices and/or cutting costs to drive additional profitability.

When evaluating its position against a strong *service provider*, a firm may consider bringing the functionality in-house via the acquisition of such providers or building the functionality internally in an effort to reduce its dependency on the constituent. An additional tactic may entail the reengineering of the firm's business process that involves the service provider in an effort to reduce or eliminate the dependency. When facing a weak service provider, the firm will likely pressure the constituent for higher value or lower prices.

Firms facing a powerful *reseller/retailer* constituency could consider a direct-to-consumer operating model, leveraging a strong brand identity. Additional tactics to get closer to the customer include improving the relationship through an improved service offering that the retailer cannot provide. Lower margins and again the direct-to-consumer model are two tactics that could prove appropriate to address less powerful intermediaries.

Establishing preferred relationships, pursuing joint marketing or branding campaigns, or bundling products/services to raise switching costs are examples of actions a firm may pursue with strong *complementors*. Alternately, weak complementors may be persuaded to allow for exclusive relationships or to open their channels of distribution to also include the firm's offerings.

When dealing with powerful *affinity groups*, a firm's plan of action may include trying to establish a strong and powerful bond with the customer itself. If developing a direct, personalized consultative relationship is not possible, the affinity group should be closely involved in the firm's planning activities in an effort to receive their endorsement and goodwill. If the affinity groups are less powerful, firms may again want to consider establishing relationships directly, or if that is not possible, at least exceed customer expectations by immediately resolving customer satisfaction issues and bringing these results to the affinity group's attention for further evaluation and communication.

If *suppliers* are in a power position, a firm's actions should focus on strengthening the relationship through collaborative planning. Other actions may include acquisitions, if appropriate, or the formation of or joining into buying groups to consolidate the buyer's bargaining position. Firms will likely address weak suppliers by squeezing their margins and attempting to prevent the constituent group from consolidating.

If a firm manages to establish relationships with strong *independent authorities*, these linkages should be strengthened via relationship building and capability sharing to ensure a lasting, favorable impression. Establishing relationships with weak independent authorities may not be a priority for most firms.

Finally, firms must address the constituency that usually is of greatest concern to them, the competition. When facing strong *competitors*, acquisitions, if appropriate, or a general improvement of its own products and services from a cost and differentiation perspective may prove worthwhile. Weak competitors should continuously be monitored in terms of both their evolving power position in the Web and their product/service offering as it is received by the marketplace.

Again, specific recommendations for action exceed the scope of this book, because such recommendations will have to be tailored to each individual organization. Yet, the Value Web framework will give you the tools with which you can begin to plot your strategic direction and tactical initiatives to establish a sustainable competitive advantage.

Summary

This concludes our introduction to Cap Gemini Ernst & Young's Value Web framework for strategy formulation. You learned that the Value Web is a customer-focused, firm-coordinated network that consists of interconnected constituents that provide elements of a total solution. Value Webs can be developed for entire industries, a specific firm located within a specific market, or for functional departments that exist within a corporation. The entity that leads the web's constituents is the one that will have established a sustainable competitive advantage. When applying the Value Web at the corporate level (versus the industry or functional department level), we use the Value Web framework to gain insights into a firm's business environment, to understand the relationships among key constituents, and to use these insights in the creation of a portfolio of strategic initiatives aimed at providing a superior customer experience and improving the firm's competitive position.

In the next chapter, we will present a Value Web that CGE&Y created for today's wireless industry.

CHAPTER

6

The Wireless Value Web

Jorn Teutloff

Introduction

In the preceding chapter, we discussed the Value Web framework as a new tool for strategic analysis. If you recall, we said that Value Webs might be constructed to map functional departments within an organization, to plot companies and their environment's constituents, or to map out entire industries. Depending on the insight that is sought by those who conduct the analysis, the Value Web framework is flexible enough to support a vast range of strategic analyses within or outside of the corporate environment. The previous chapter illustrated the approach using a fictitious company, Acme Financial. Acme was presented as an example of a firm that performed the analysis to better understand its immediate playing field. Taking the approach to an industry level, this chapter will walk you though a Value Web that CGE&Y prepared for the general mobile/wireless landscape in an effort to reveal the industry's various segments and the types of companies represented in each.

The Wireless Value Web's Origin

The work we are presenting in this chapter was prepared by the CGE&Y's Mobile Commerce Center of Excellence. The Center maintains a staff of dedicated resources who continually monitor industries and technologies, including mobile and wireless, by identifying, investigating, and mapping companies, their offerings, and their movement within the space. Because the mobile technologies industry is relatively young, new players appear on a monthly basis, established companies frequently shift direction, and unprofitable ones disappear seemingly over night. Unless watched closely, the mobility landscape, and more importantly, any associated opportunities, can be difficult to pin down. What looks like an attractive niche today might be a dead end tomorrow. Thus, the purpose of the Mobile Commerce Center of Excellence is to always be fully aware of the latest seismic shifts so that the firm can offer its client base access to up-to-the-minute research and insights about the advances of the rapidly evolving field.

The Wireless Value Web is frequently being leveraged as a starting point when strategic plans and tactical courses of action are created for CGE&Y's clients. Before attempting to set a general course of direction or craft specific initiatives for a client, the team must be clear about who is who in today's mobile technologies space. Once the playing field is clear, the team can proceed with developing a mobile/wireless strategy and associated tactical initiatives, whether the client is a new company about to enter the arena, or an established organization that seeks to leverage these new technologies within its current operations.

Today's Wireless Value Web

As of the writing of this book, the Wireless Value Web we present here reflects the industry's status quo. There is no doubt in our minds that some of the companies mentioned in the following sections will have ceased to exist, whether they were acquired, merged, or simply went out of business by the time this book reaches your desk. Nevertheless, the general concept of a web of companies servicing the mobile technologies market space will likely remain in effect for some time to come—even if specific companies come and go. Before we present you with some of the companies that mark today's landscape, we need to briefly introduce a framework of organizing these companies.

Taxonomy

When we developed Acme Financials' Value Web in the previous chapter, we identified the following constituencies:

- *Customers* are at the center of the web.
- The *firm* provides the offer a customer considers to purchase.
- *Competitors* are vying for the same customer.
- *Service providers* offer enablers that are required to conduct the transaction.
- *Suppliers* have provided inputs that went into the firm's offering.
- *Resellers/retailers* provide a channel and other value-added services.
- *Complementors* offer goods and services that add to the usability of the firm's products.
- *Independent authorities* include the media and governmental regulators.
- *Affinity groups* consist of the customer's family members, friends, or business associates.

The set of constituents provided a good taxonomy for us to describe Acme's—or any other firm's—universe and to assist with crafting the direction required to enhance the offering a firm provides to its customer segments. However, when we expand our analysis from a company to an industry level, we need to introduce a slightly different categorization format. The adjusted taxonomy we are going to apply to our analysis of the mobile technologies industry steps away from the generic constituencies identified within Acme's universe and introduces an additional set of groupings that is unique to the industry under investigation. Whereas the constituencies we described in the previous chapter made sense from the perspective of the customer who considers purchasing an offering put forth by a firm, they no longer make sense when our field of vision opens up to encompass an entire industry. For example, it was quite obvious for the competitor constituency to have a solid presence in Acme's Value Web because Acme's customers would sooner or later be exposed to these companies. In other words, because we were dealing with a transaction-oriented relationship between a customer and Acme—the firm providing the offering the customer is interested in—it was easy to identify those companies that offered similar products and, by definition, group them into what we called the *competitor* constituency.

When we investigate entire industries, however, a precise customer-merchant relationship is not yet defined, thus rendering the category of "competitor" for the purposes of our analysis meaningless. Who is the customer, and who is the offering firm when looking at the entire industry established around mobile/wireless technologies? Is it the end-consumer who is interested in purchasing a wireless device from a hardware manufacturer? Or is it the consumer who already owns the device and is now looking to purchase airtime from a telecommunications carrier to activate the device? Within a generic industry analysis, such as that of the wireless industry, everyone is a competitor—vying for some customer's attention. Hardware providers, network operators, content providers—all are pursuing the end-user. Only when a transactional relationship is about to be formed between a customer and the firm—which we previously defined as the company offering the preferred solution to a customer's unmeet needs—do perspectives fall in place. Now, with a stake in the ground, a reference point established, we can begin to group individual companies into constituencies. Thus, only after we have identified the firm that a customer intends to purchase from can we determine which entities would fall into the constituencies of competitors, complementors, and so on.

Another way to explain the need for a new taxonomy is to look at how companies are classified within a Value Web. In the case of the previous chapter's analysis, we defined the groups of constituents in relation to the offering that was presented to the customer. A competitor would attempt to replace the original offer with its own, a complementor would enhance the offer, affinity groups would provide information about the offer, and so forth. Within an industry analysis, however, we are looking at a large variety of dissimilar offerings that target different customers. Thus, we need to find a better, more comprehensive structure for grouping companies that takes into account the bigger picture.

This new structure, or taxonomy, that we will apply throughout the remainder of this chapter is illustrated in Figure 6.1. At the highest level, there is the industry under investigation, followed by multiple segments of companies within that industry, multiple categories of entities within each segment, multiple groups of entities within each category, multiple types within each group, and multiple constituencies for each type

At the *industry* level we are specifying simply which industry we are investigating. In our case this is the mobile technologies industry, but it could as well be the automotive, insurance, or healthcare industry. The next level covers industry *segments*. Here, we are getting somewhat specific to the industry our analysis applies to. For example, some of the segments within the mobile technologies industry we will describe include user devices, content, or technology enablers. Within segments we find *categories.* Within the user devices segment, for example, are the portables and peripherals categories. Peeling the next layer of the onion, we encounter *groups.* Examples of two groups within the portables category are cell phones and PDAs. Within the PDA group, we are looking at various *types,* such as Pocket PC or Palm OS–powered PDAs. Lastly, we move to the final layer, our familiar *constituencies.* To close the loop, within the Pocket PC type of PDAs there are *firms,* which offer the hardware a customer is considering to purchase, *competitors* that provide similar products, and so forth. Examples of firms offering the hardware include Compaq, among others. The complete path can be expressed as follows: *Industry: Wireless Technologies/Segment: User Devices/Category: Portables/Group: PDAs/Type: Pocket PC/Constituency: Firm: HP.*

As a final note about the new taxonomy, the mobile/wireless industry landscape we are presenting here uses industry-specific terminology within the generic taxonomy presented in Figure 6.1. For example, we shortly will be talking about "content aggregators" and "network operators" as some of the categories of entities we will describe. Although this type of industry jargon makes sense for our analysis of the wireless web, it would obviously not apply were we to talk about the automotive industry, for example.

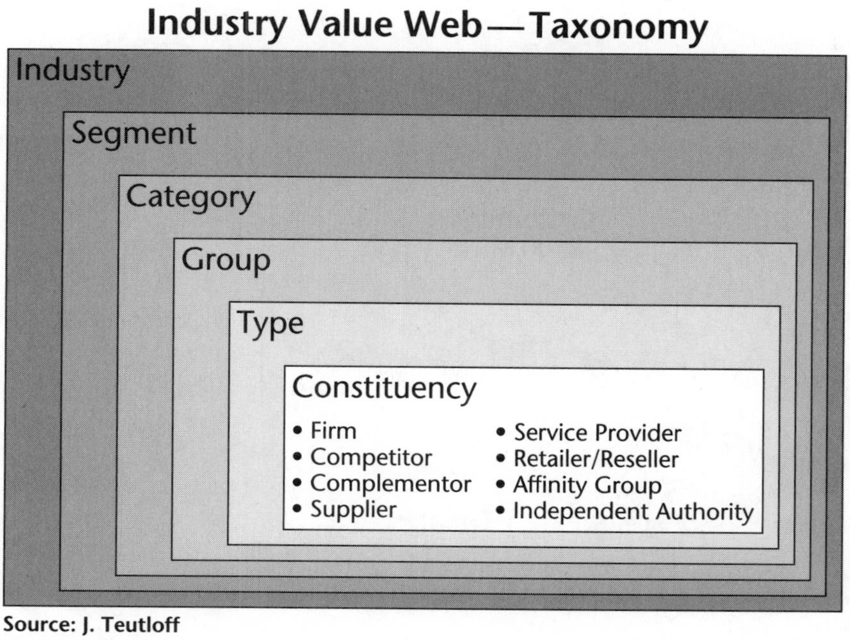

Source: J. Teutloff

Figure 6.1 Industry taxonomy.

Figure 6.2 illustrates the industry segments of our high-level Wireless Value Web. Notice the layers, or orbits, that form around the core. Located at the center of the web are customers/end-users. The first layer around the center includes three segments: user devices, content, and connectivity providers. Forming the second layer are technology enablers. The fourth layer shows environmental enablers. We will discuss each segment in detail. The reason we use a slightly different graphical representation of our web than we did for Acme is that the new layout, showing companies located on several orbits around a core, illustrates the additional dimension of a vendor's product proximity to the end-user. Vendors located on the first orbit provide offerings that the customer consciously interacts with, such as, for example, a cellular phone. Companies on the outer orbits provide supporting products and services that make the technology work, but that are usually not touched by a customer.

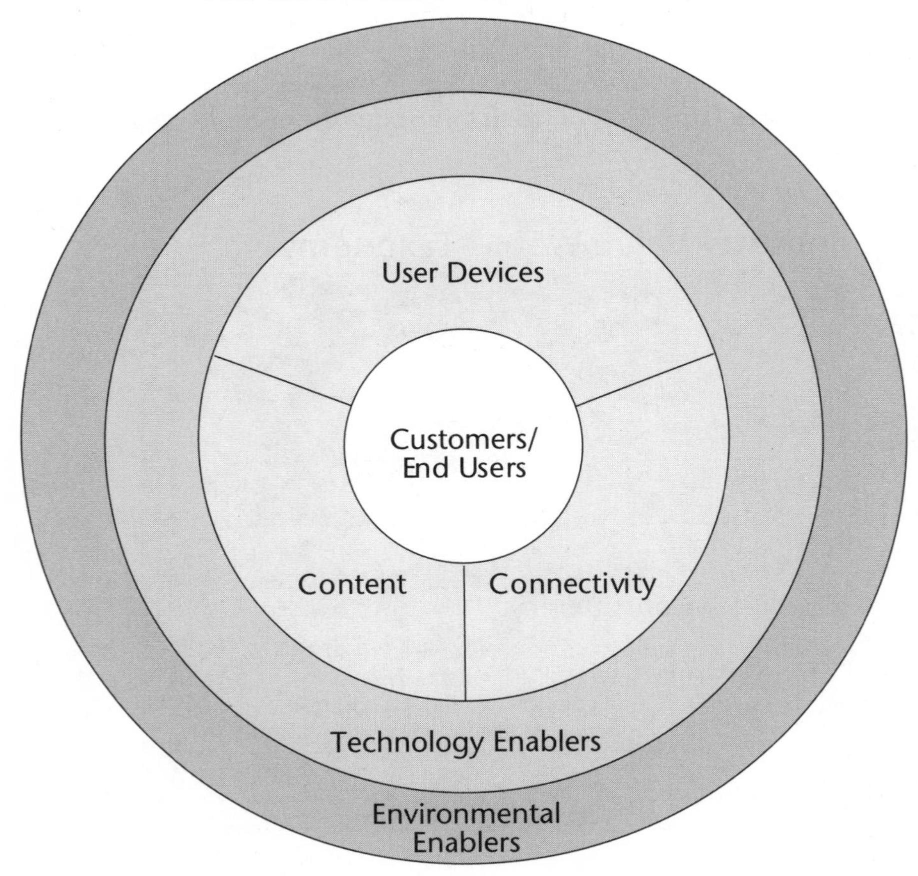

Source: CGE&Y, J. Teutloff

Figure 6.2 The generic Wireless Value Web.

Industry Segments

As mentioned, the Wireless Value Web we are describing in this chapter is an attempt at packing an entire industry into a structure that facilitates our understanding of the industry's current state. The following pages will present a snapshot of the wireless industry as of the writing of this book to provide you with an overview of the market space. Understanding the playing field is the prerequisite to crafting your own strategy and tactics for how you and your company can take advantage of the opportunities presented by mobile and/or wireless technologies. To accomplish this feat, we will introduce and discuss the following industry segments and associated categories.

- Customers, also called end-users
- User devices
  - Portables
  - Embedded devices
  - Peripherals
- Content providers
  - Content creators
  - Content aggregators/portals
  - Application developers
- Connectivity providers
  - Network operators
  - Wireless service providers
  - Hosting companies
- Technology enablers
  - Enabling hardware
  - Enabling software
  - Professional services
- Environmental enablers
  - Research firms
  - Financial services providers
  - Trade associations
  - Standards groups
  - Regulators

In addition to these industry segments and categories, we will explore the groups and types of entities located within each. Furthermore, to relate our theoretical framework to the real world, we will also list some of today's companies—most of them U.S.-based—within each type. Please note that we neither claim that the listings are exhaustive, that is, include all companies in a given space, nor endorse any of them or their services. We merely cite these providers as examples for the benefit of those readers who would like to learn more about any given type of functionality provider. Also, we should point out that our clustering of product and service providers into segments, categories, and the like is not necessarily mutually exclusive; the reader will notice a certain overlap, especially with those companies that offer multiple wireless products or services. Again, by the time you read this, there will be new contenders in the wireless area, whereas others may have disappeared into obscurity. In other words, exercise caution when using the lists and do your own research to confirm your marketplace.

Customers/End-Users

Let's start with the core of the diagram. Figure 6.3 illustrates the customer/end-user segment as the center of the Wireless Value Web and presents the segment's four categories: residential customers, businesses, educational institutions, and governmental organizations.

Customers, in our text, also referred to as end-users even though they are not necessarily one and the same, are located at the center of the construct because, as you recall, the concept of the Value Web stipulates that any analysis has as its ultimate goal to enhance the offering(s) provided to the end-user. In the mobile/wireless space, the end-user is any individual or organization who consumes content provided via a handheld device for reasons of communication, information acquisition, education, entertainment, or commerce.

Residential

Residential customers encompass individual end-users such as you and me, who are using our cell phones to conduct a telephone conversation, engage in a stock trade, play games while being delayed at the airport, or consult our wirelessly-enabled PDA for driving directions in an unknown part of town. The aggregation of individuals who use mobile technologies for personal communication, information acquisition, education, entertainment, or commercial purposes constitutes the mass market that many of the providers of hardware, content, and connectivity are pursuing. Over the last few years, the world has witnessed a skyrocketing penetration of wireless devices (and associated network usage) at the consumer level.

Cited frequently as the shining example for the power of wireless, NTT DoCoMo's iMode service in Japan illustrates how wireless services can constitute an immense business proposition for companies that manage to capture the mass market. Launched in February of 1999, the service catapulted to over 40 million subscribers by quarter 1 of 2003; a wildfire-like growth fueled by inexpensive handsets, always-on connectivity, and a plethora of data services—accounting for about 80 percent of NTT DoCoMo's revenue—provided by over 50,000 commercial content providers. In Europe, especially in the Nordic countries, mobile phone penetration levels have reached between 70 and 80 percent of the population, whereas the U.S. market is still playing catch-up with approximately 140 million subscribers, or about 49 percent of the population, by the end of 2002, according to CTIA.

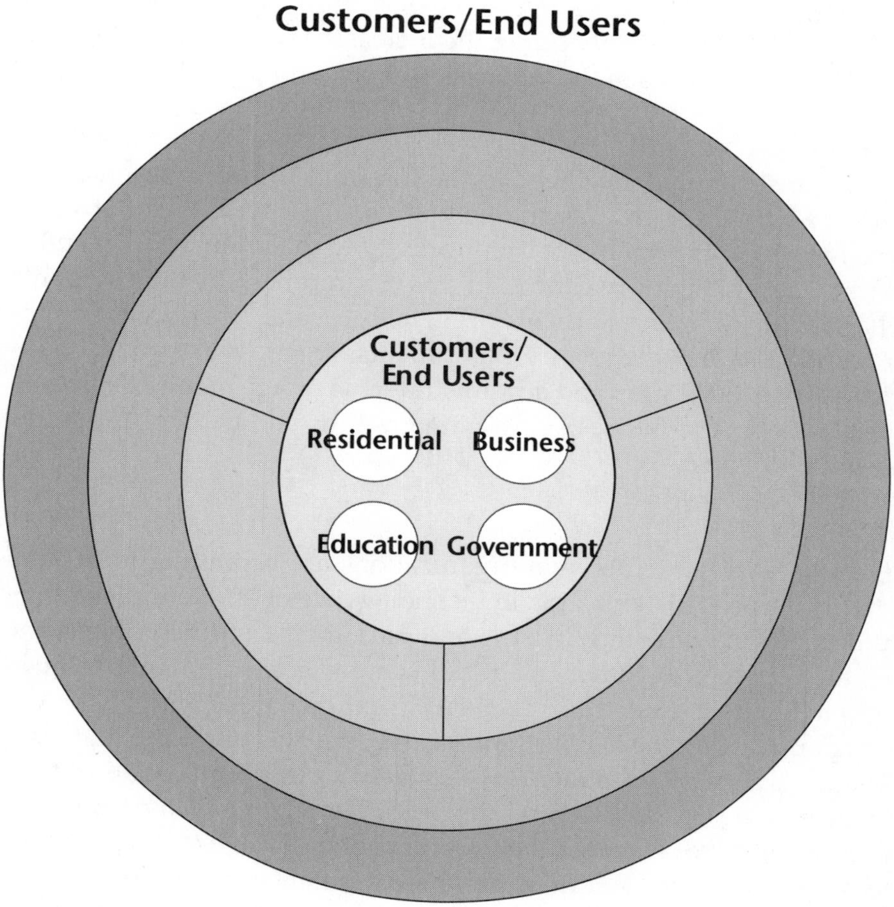

Source: CGE&Y Analysis

Figure 6.3 End-users are located at the center of the construct.

Business

Businesses are a second category of end-users within the customer segment. Companies and their employees as well as all other professionals who use mobile/wireless technologies for business purposes fall into this category. Business users use this new technology for a wide range of applications, which include communicating with colleagues, suppliers, or customers, accessing corporate e-mail, engaging in personal information management such as calendaring or time and expense submissions, gaining access to back-office ERP systems, receiving the latest service dispatch, closing a deal at a customer site, or prescribing a medication for a patient during a house call. According to Gartner Group, the job requirements of over 25 percent of the 112 million members strong U.S. workforce entail a mobile component. In a 2002 study published by Cahners In-Stat Group, the research company forecasts telecom expenditures to grow by roughly 8 percent through 2005, emphasizing the increasing penetration of mobile and wireless technologies in the business sector.

Education

Educational institutions have been at the forefront of deploying wireless technologies by equipping classrooms with wireless local area networks that provide students with instant connectivity to the Internet upon entering the facilities. Other applications in this environment include students using handhelds to maintain their class schedules and homework assignments, record and analyze science experiments, or access electronic books and dictionaries. Educators and administrators leverage mobile devices to manage classroom attendance, assist with grading, and maintain appointments and to-do lists.

Government

Last, but not least, governmental organizations are beginning to explore the benefits associated with mobile technology devices. Examples range from police officers querying records and files and recording crime scene data, all the way to governmental entities using wireless networks to push messages to handset users during times of a national emergency.

Moving outward in the Wireless Value Web, we encounter the first orbit that surrounds the end-user. Figure 6.4 shows that this orbit contains three industry segments and associated categories. The segments located in this first layer around the center are the providers of user devices, content, and connectivity. Each entity within these three segments offers a product or service that involves the end-user, usually in a very direct way, meaning that end-users have direct access to these products and services. Later in this chapter, we will encounter segments located in layers farther away from the core. Companies in these orbits usually do not interact with the customer, but instead deal with the companies located in the previous layer.

On the first orbit around the end-user core of the web, we find manufacturers of user devices, content providers, and connectivity providers.

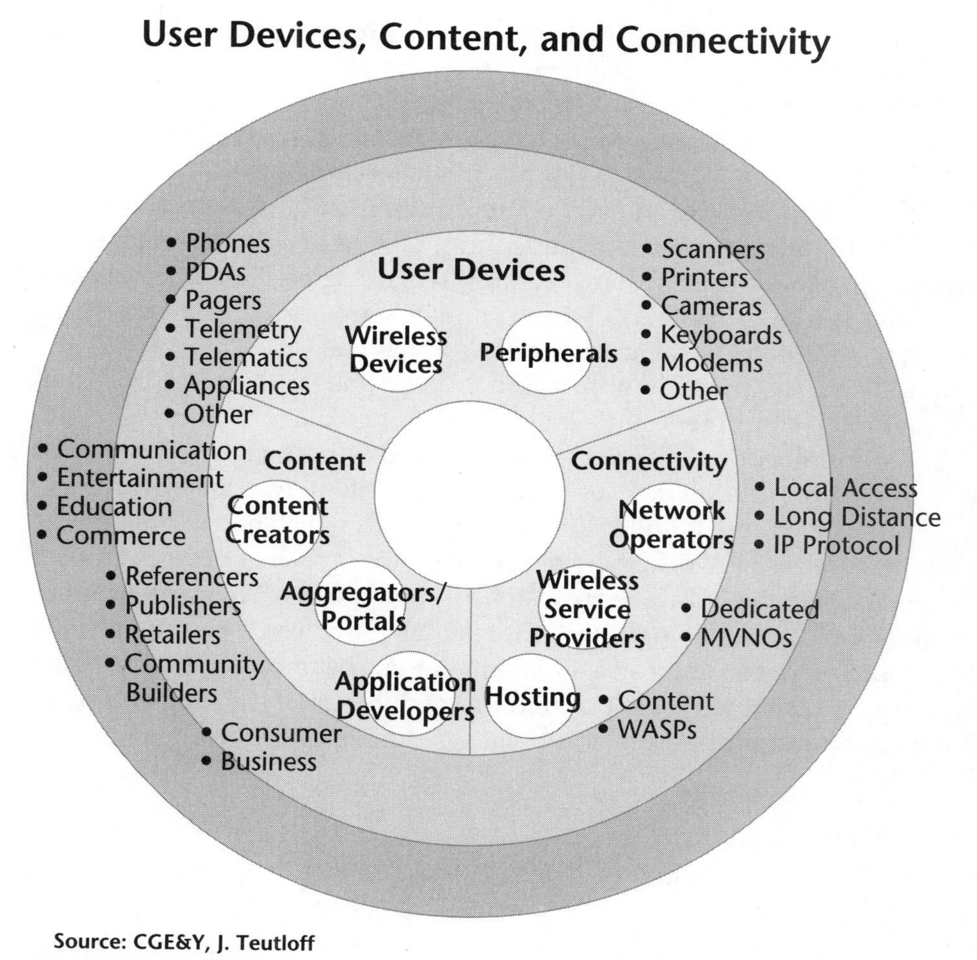

Figure 6.4 User devices, content, and connectivity.

User Devices

The user devices industry segment entails categories of companies supplying consumers with portable devices, embedded technologies, and peripherals.

Portables

Manufacturers and distributors of handsets, including cellular phones, PDAs, smartphones, two-way pagers, telemetry and Telematics devices, wireless Internet appliances and other hybrids offer the gateways through which end-users communicate, access content, or conduct transactions. Companies that are well known for their wide range of data-enabled cellular phones and so-called smartphones, the combination of a cellular phone and a PDA, include Ericsson, Handspring, Kyocera, Motorola, Nokia, Panasonic, Samsung, Sanyo, Sony-Ericsson, Siemens, Touchpoint/LG, and others.

PDAs are supplied by an even larger group, including Acer, Audiovox, Casio, Dell, Fujitsu/Siemens, Handspring, Hewlett-Packard, NEC, Palm, Research in Motion, Sony, Sharp, Symbol Technologies, T-Mobile, Toshiba, Viewsonic, Zayo, and others. Two-way pagers are currently marketed by Glenayre, Motorola, Research in Motion, and Skytel.

In addition to "pure" handhelds such as phones and PDAs, portables such as laptop computers and tablet PCs fit in this category. Equipped with add-on (or embedded) wireless network cards, these computers can be used as communication devices or to receive content over the airwaves.

Besides these more familiar interfaces through which end-users interact with the wireless network, there are several other types of devices that are on the market, including wristwatches powered by Microsoft's Smart Personal Object Technology (SPOT), telemetry devices for automatically indicating or recording measurements at a distance, and Telematics devices that combine global positioning satellite (GPS) tracking and other wireless communications for automatic roadside assistance and remote diagnostics.

Embedded Devices

Especially when it comes to GPS-equipped devices, the lines between portables and embedded devices gets blurry; mapping technology is available both as a stand-alone handheld, or as a technology feature integrated with your automobile's navigation system. Furthermore, we are beginning to see wirelessly enhanced appliances—ranging from networked refrigerators to microwave ovens and washing machines that have wireless capabilities embedded. Surely, we are just beginning to see the many uses of wireless technologies at the device level. Last, there are other embedded devices that users rarely interact with, such as wireless payment gateways that are built into vending machines and transmit inventory and operations data to the machine's owner or operator.

What are some of the strategic courses of action device manufacturers have at their disposal? First, manufacturers must focus on making their appliances easier to use while adding features and functionality that further drive adoption. Especially in the United States, wireless devices have not yet reached mass-market status which is partially due to the incompatibility of competing carrier network technologies that, for example, until very recently prevented users of one network from sending short messages to subscribers of another. Another reason for the slow adoption of data services include confusing rate plans, some of which offer a flat billing per month, whereas others charge by the megabyte of data transferred. How is the average consumer to know how large—in kilobytes, mind you—a text file or a digital photo is? As long as consumers have to evaluate a multitude of competing billing plans for voice, and now also for data services, confusion will curb uptake.

A potential strategy for manufacturers to increase sales entails the segmentation of the market and developing devices that appeal to each segment through their special form factor and/or applications. Examples for a form factor segmentation approach include those manufacturers that create "ruggedized" versions of their PDAs for use in, for example, industrial environments. Other device providers partner with application developers to deploy devices that are specialized in function. For example, in the past we have seen cobranded, two-way pagers that provided Microsoft Hotmail users with access to their Hotmail account, and their account only. Similar alliances targeted users of Yahoo or AOL e-mail and instant messenger functions by providing devices that are limited in functionality to access these service providers. Although some of these differentiations may not have been too successful, especially when limiting consumers regarding e-mail access, they illustrate the important concept of personalization versus a one-size-fits-all approach. Business users in particular are leading the way in terms of individualized devices that are designed for specific purposes.

Once adoption rates have plateaued, device manufactures must think about how to encourage hardware upgrades and/or replacements. Beefing up cell phones with new features such as chips with ever-increasing memory to capture entire phone books and calendars to changing appearances such as device dimensions and screen size are popular tactics to keep devices rotating through the hands of an evolving user base. In addition, recent enhancements to wireless data networks have provided us with 2.5G, and in some areas 3G, bandwidth. Being able to carry more data at faster speeds, these networks have spawned the addition of small digital cameras to the cell phone. PDAs, on the other hand, are gradually being equipped with internal or external 802.11 wireless LAN modules. Speaking of peripherals, let's go on to the next section.

Peripherals

Whereas cellular phones can be equipped with snap-on digital cameras, keyboards, or battery chargers, it's the more versatile PDA that enjoys a plethora of optional devices that complement the core product. Adding to the functionality of PDAs are peripherals such as headsets, keyboards, scanners, and printers. Other accessories include cellular phone modules, radio tuners, game and reference modules, and digital cameras that plug into the extension module slot of the device. More esoteric peripherals include text-to-speech decoders that enable your PDA to "read" downloaded books, magazines, or newspapers to you while you exercise or drive to the office. For the hiker, sailor, or urban street warrior, there are GPS modules that can be added to your handheld PDA and thus render it a navigational aid, promising that you'll never get lost again. Another important group of companies in this category includes the manufacturers of wireless modems in the form of expansion packs or modem cards that allow the device to access data via the Internet. Without these cards, the handheld device lacks the connectivity to the Net—connectivity without which the PDA loses much of its appeal for the mobile user. Last, there are a slew of companies supplying battery packs, memory cards, carrying cases, belt clips, styluses, automobile adapters, synchronization cradles, and other gadgets that add to the usability of the device.

The critical success factors for peripherals providers are closely aligned with the strategies and tactics deployed by the device manufacturers, which can be a double-edged sword. On one hand, such a close alignment allows these companies to sit back and observe what direction device manufacturers take and adjust their own course of action accordingly. Observing where the industry's 800-pound gorillas are heading illustrates a follower mentality that may allow a peripherals provider to avoid costly mistakes associated with venturing forth into uncharted territory on their own. On the other hand, that close alignment also means that peripheral providers depend to a large degree on the success of the manufacturers—for which there is no guarantee in an ever-evolving marketplace. New products may be short-lived or not successful at all, causing peripheral providers to be shrewd about timing the release of their offerings. By jumping onto a bandwagon too soon, they might find themselves having invested scarce resources in a train that is headed nowhere. If they jump on it too late, there might be several other players already on the field, leveraging their first mover advantage to form exclusive alliances with the device companies, which effectively keeps other competitors at bay.

Content

After our brief exploration of device manufacturers, we are ready to continue to content providers. Content providers are companies that provide the information and applications the end-user is accessing via the handheld device. As such, content providers can be categorized into content creators, aggregators/portals, and application developers.

Content Creators

Before we talk about *content creators*, let's quickly think about how we could organize our discussion around content. There are two major content categorization approaches that will allow us to structure our discussion: content by form and content by function. First, when thinking about content and its form or format, we can distinguish between three areas: written, audible, and visual. Written content includes books, journals, poems, newswire reports, newspapers, and the like. Visual content consists of video, photographs, drawings, and paintings to name the most obvious. Audible content entails songs, speeches, and music. Of course, the three dimensions overlap; magazines are a good example of written material with a highly visual slant; movies and theater productions exemplify the combination of visual and audible content; books are frequently available in both the traditional format and on tape or CD-ROM to be listened to. At the intersection of all three dimensions, we may find television commercials or multimedia-enhanced Web sites, heavily combining all three content formats.

On the other hand, we could discuss content along the lines of functionality. Content might be created for communication, information, entertainment, education, or commerce. Again, the segments are partially overlapping. For example, whereas online multiplayer games are exclusively entertainment-oriented, and online reference materials clearly aim at education, the applications category of "edutainment" illustrates how entertainment can blends with education in an effort to make learning fun. Similarly, whereas the short message system was originally perceived to allow cell phone users to communicate via short, text-based messages created via the phone's keypad, the technology is now also being used for entertainment purposes in the form of text-based games, and even commerce in the form of push advertising.

The point is that any discussion of content in the Wireless Value Web could be organized by form or function. For our industry analysis, we will use the latter approach, that is, we will look at the content creation market in terms of features and functionality.

Let's begin. As proposed, the groups of entities within the content creators category include those that manufacture content for communication, entertainment, education, and commercial purposes disseminated via mobile/wireless devices. Content for each purpose is available from major carriers, entertainment companies and a plethora of small, entrepreneurial enterprises. Content includes text, music, video, downloadable ring-tones, and animated icons that appear on cell-phone's screen created by a bevy of artists. Other originators of content are companies that gather data and transform data into information, for example, in the form of reports about the latest weather or road conditions. Content is offered to the customer directly from the source or via the wireless carrier that the end-user is subscribing to. If accessed directly, the content is made available for downloading, via snap-on modules, or in real-time over wireless networks. If procured from the carrier, the content is usually accessed via the carrier's wireless portal.

Another group of content creators that needs to be mentioned are merchants who make their products or services available via wireless networks. Such merchants range from retailers of branded apparel, consumer electronics, travel, hospitality, and entertainment products all the way to restaurants or financial services institutions, including brokerages, insurance companies, and banking services. The list is seemingly endless. Think of the many brands that are selling directly via the Internet. Even though we are still a long way from recreating on the PDA or cell phone the rich Web experience we are used to when sitting behind a PC, all the content available to us on the Net will eventually make its way to the handheld gadget.

One last comment regarding content creation. Although third parties, such as artists and information providers, are the most obvious creators of content, an even larger number of content creators are end-users themselves. With the advent of the Internet, you and I have become creators of content available to millions of other browsers, all around the globe. Whenever we write an e-mail letter, fill out an online feedback form, rate a consumer product we have purchased, create an online photo album, or develop our own Web pages, we are creating electronic content that sits on a server somewhere, waiting to be accessed by someone. And while the origination of own content and sharing it with others is at this time mostly limited to the desktop computer system, it won't be long until advances in wireless technology—especially in bandwidth and display size—will allow us to move these capabilities onto handheld devices. Content creation has been expanded from being an activity previously solely performed by trained professionals to a hobby taken up by the masses.

In their quest for profitability, content creators must create personalized content that is unique, popular, and convinces end-users to pay for it. For customers on the go, such content means information that contains an element of criticality, accessed via a handheld device. Critical information that is attached with a monetary value is information that end-users need immediately, wherever they are. Different for each user, examples of such information include real-time stock quotes, location-aware driving directions, the latest sports scores, and customized alerts that indicate when a special product you placed on your virtual wish list goes on sale at the local department store.

Content Aggregators

When information created by a source is not directly delivered to the end-user, but is combined with other content before being published, we are talking about the services *content aggregators/portals* provide. Within this category of content providers, we can further distinguish between referencers, publishers, retailers, and community builders.

In the Internet environment, aggregators that provide reference services organize data and sites into easily searchable areas. Generating revenue mostly from advertising, these companies often act as doorways, or portals, to the Web, helping consumers find information. Referencers such as the Internet portals AOL, MSN.com, Citysearch, and Yahoo are well known for their expertise in packaging and marketing data. Instead of requiring you to be aware of individual content creators and their Web sites, your friendly content aggregator has done all the work for you and provides you with an exhaustive package of information, culled from multiple sources. If you don't like what you see, the aggregator provides a search engine for you that brings back even more information. Internet portals maintain their members' profiles and deliver aggregated information, frequently personalized, right to the member's virtual doorstep.

In the wireless space, the concept is the same. Instead of having to search for information by guessing Web site addresses of content creators, end-users simply use an aggregator that has organized the information in an easy-to-use format. In the United States, the larger wireless portals include AOL Wireless, AvantGo, MSN Mobile, Palm.net, and Yahoo Mobile. In

Europe, BT's Genie, Iobox, Mviva, and Vizzavi provide popular portals, whereas J-Sky and NTTDoCoMo are the strongest contenders in Asia. Wireless portals increasingly provide to the users of handhelds some of the same content aggregation services as full-blown, traditional Internet Web sites. In addition, wireless carriers such as Alltel, ATT Wireless, Cingular, Nextel, Sprint, T-Mobile, and Verizon have established basic wireless portal services for their customer bases.

One more word about referencers that act as portals. The two types of portals we distinguish between are vertical and horizontal portals. Personal finance sites, such as Quicken Mobile, CBS Marketwatch Mobile, and Yahoo Finance are examples of portals that offer a deep experience with highly relevant content relating to the subject at hand. On the other end of the spectrum, we find horizontal portals, operating sites that accumulate content from seemingly any walk of life. Yahoo Mobile, MSN Mobile, and other portal operators belong in the group of referencers whose goal is to become the thoroughfare through which the user accesses whatever information he or she is seeking.

The second group within the content aggregator cluster encompasses publishers. Basing their revenue on advertising, subscriptions, and commissions, publishers aggregate news, travel products, or other types of information for consumers to access. Examples for companies in this space include CNN, the *Financial Times*, MSNBC, Microsoft Expedia, Travelocity, the *Wall Street Journal*, and others. Many publishers maintain archives of the information presented, allowing end-users to search such archives at a later date.

Retailers in the aggregator category aggregate merchandise or services from many different companies and make them available for sale in an online supermarket type of format. The retailers we recently mentioned within the content creator category are merchants selling their own brands, and their brands only. Retailers that belong to the content aggregator category offer merchandise within a mall context, in which multiple brands are featured in their own virtual locations, or grouped by function, regardless of brand. Revenues are mostly derived from sales commissions. For example, although we would consider hpshopping.com a *content creator* in the consumer electronics space, the company's products are also available from online *content aggregators* such as the online retailer buy.com.

Last, we consider community builders as a unique group within the content aggregator category. Community builders foster the creation of network communities around specific topics of end-user interest, such as hobbies, geographical location, professional interest, or self-help issues.

For content aggregators, the name of the game is critical mass. Unless we are talking about aggregators that are representing offline brands and that are frequently backed by their deep-pocketed parents, wireless portal operators must hustle to establish brand recognition. Their foremost concern is to quickly attract and retain a large user base in an effort to attain profitability from a revenue model that is built upon advertising fees of referral/sales commissions. Yet, creating "sticky" portals is no small feat; entry barriers are low, and the competition—as the short history of the Internet has shown—is fierce. Those players who will be successful in the wireless portal game are the ones who establish alliances with the major content providers (upstream sourcing) and carriers and wireless service providers (downstream distribution) to tap into fresh content from the former, and an already existing customer base from the latter.

Application Developers

The third category of content providers in our Wireless Value Web includes *application developers*. We are calling out this group although one could easily make the case for including these software programmers in the content creator category. Yet, if we look at the type of content produced by developers, it becomes clear that these individuals or companies operate in a league of their own. What separates applications from content such as a news story, an image, or a physical product available for sale on a Web site is the complexity of the application as measured in the time it takes to develop the content combined with the breadth of functionality the end-user receives by interacting with the application using the mobile device.

Application developers, then, write software programs that the end-user directly interacts with on the cell phone or PDA. In this definition, application developers create computer programs that are geared towards the consumer, as for example, entertainment software, educational software, or personal information management applications. Similarly, this category of content providers develops business software. These applications are used in the corporate environment to allow direct access into a company's back-end system, pulling and/or pushing financial data, inventory levels, customer information, or other operations-related data onto a mobile device. Vertical applications appear on the market at an accelerating pace, including solutions for customer relationship management, sales force automation and effectiveness, field service, warehousing, transportation and logistics, and IT management, for example. Horizontal applications, as the name suggests, cover a broad spectrum of business functionality such as personal information management (PIM), time and expense reporting, calendaring, or appointment scheduling across multiple sectors.

Additional examples of business functionality provided by application developers include access to the corporate e-mail systems for employees on the go, access to inventory status or price lists for mobile sales representatives, and real-time visibility into the availability of parts and customer transaction histories for service technicians in the field. There are simply too many companies in both the vertical and horizontal applications categories to list here, but suffice it to say that we have only just begun exploring all the software wireless devices will run in the near future. By the way, companies that provide enabling software, such as transaction engines that allow wireless users to securely pay for merchandise with a credit card, are described further when we examine technology enablers.

Application developers provide content that is much more complex than an up-to-the-minute weather report or the latest financial news headlines. Developers are aiming to create personalized applications that capture customer appreciation and loyalty. They must innovate early and differentiate their offerings in lock step with the rapidly evolving wireless device landscape. Applications that are included with a device at the point of sale usually have a better shot at reaching critical mass. Nevertheless, nonaffiliated application developers whose products must be purchased on the after market may develop strategic alliances—be it with device manufacturers or content aggregators—that can provide them with preferred-vendor status. Closely aligning with a big player can improve the developer's chances of gaining brand recognition and market share.

Connectivity

The last segment of entities located within the first orbit around the end-user is the connectivity provider. These are the companies that provide various types of wireless network access services through which a customer connects with another, or receives content. In return for access, consumers pay a monthly service charge, which allows metered access (that is by the minute, or by the amount of data transferred) or unmetered (all-you-can-eat) wireless traffic. Other payment options include prepaid or pay-as-you-go. Connectivity providers include network operators, wireless service providers, and hosting companies.

Network Operators

The first category within this segment entails *network operators*. In general, this category entails local access providers, long-distance network operators, and Internet Protocol (IP) transport providers. Local access providers own and operate the physical networks that connect end-users to an Internet service provider's (ISP) points of presence (POP), local exchange carrier switches and cable television *headends* (facilities that originate and communicate cable modem and cable TV services to subscribers). Types of local access providers include local exchange telephone network operators, cable television network operators, and—in the Wireless Value Web—mobile telephone network operators. Long-distance network operators own and/or operate physical trunk networks on which various forms of telecommunications travel between localities and countries. Types of companies within this group include long-distance telephone network operators, satellite operators, and operators of terrestrial microwave networks. IP transport providers operate the networks that carry Internet protocol traffic between POPs and resell Internet transport services to end-users. ISPs and Internet backbone operators fall into this group.

Examples of network operators include companies such as ATT Wireless, Alltel, BT Cellnet, Cingular, Deutsche Telekom, Nextel, NTT DoCoMo, Orange, Sprint, TeliaSonera, Telstra, Telus, T-Mobile, Verizon, Vodafone and many, many others around the globe.

Essentially, network operators provide the voice and/or data services that enable a wireless device to function. In addition to connecting their large customer bases to other customers and content, network operators maintain the contract and billing relationship with the end-user, thus controlling a very important customer touch point. Critical success factors on the front-end include for network operators to develop or secure innovative end-user applications. Constant innovation can provide the operator with a first mover advantage that provides for differentiation from the intense competition. Especially as average revenue per user (ARPU) from voice communications falls in relation to revenue from data services, competition for end-users is steadily increasing. On the back-end, network operators must implement state-of-the-art self-service operating and business support systems that allow for the efficient management of millions of customers.

Wireless Service Providers

Providing access services to end-users similar to those offered by network operators are *wireless service providers*. Wireless service providers focus exclusively on servicing the wireless community, while the previously discussed network operators have their roots in traditional voice communication. There are a few wireless service providers that build and maintain their own proprietary networks, but most purchase excess capacity from the big network operators and resell the service under their own brand. For example, the now defunct Ricochet service by Metricom was a proprietary high-speed data network that offered wireless connectivity for PDAs throughout select major U.S. cities. On the other hand, Virgin Mobile, the UK-based wireless service provider and part of the Virgin Group, is an example for a Mobile Virtual Network Operator (MVNO). Currently on an expansion path within the U.S., Virgin Mobile has formed an alliance with Sprint from which it obtains excess capacity. Virgin Mobile brands the service using the Virgin trademark, and applies the Virgin Group's strong branding and superior marketing expertise to sell the offering to the youthful community that the company has established close ties with through its worldwide music empire. In addition to Metricom and Virgin Mobile, other companies currently operating within the United States include GoAmerica, Earthlink, Motient, and Palm.net—all offering the wireless data service that connects PDAs to the Internet.

Wireless service providers face strategic challenges similar to those of the previously discussed network operators when it comes to gaining a differentiation advantage and efficiently administering large customer bases. For those companies building and maintaining their own networks, access to capital and being able to quickly establish critical mass is most important to satisfy the immense startup costs associated with running a proprietary network. Resellers such as MVNOs not only must be experts at forging long-term relationships with the carriers they purchase capacity from, but also must have brands that are strong enough to extend into the telecommunications market.

Hosting Companies

The last category we want to briefly touch upon contains *hosting companies*. Within this category, we find companies that store all the content. Too many to list, content hosters provide a valuable link in the interaction between content users and content creators/aggregators in that they operate and maintain the servers on which the material is physically stored and processed, usually in return for a monthly or annual hosting fee.

Wireless application service providers, or WASPs, provide another category of hosting service in that they not only store content, but also more importantly run applications that process such data. WASPs perform the same services as their wireline brethren in that they host, manage, and maintain business or consumer applications on an outsourced basis. Yet, WASPs add the mobile dimension, because the managed applications are accessible via mobile/wireless devices. The benefits to a company of securing services from a WASP versus creating them in-house include faster time to market, reduced operating expenses, avoiding the acquisition of new hardware, and not having to build a staff of IT professionals versed in still-evolving mobile technologies. Drawbacks, besides the familiar issues surrounding wireless access coverage and bandwidth, both of which should be mitigated as new technologies are deployed over the coming years, include questions about handing sensitive data to a third party, the lack of being able to customize the application to fit the outsourcer's requirements, and the market's still being in its early stages with no clearly established leaders. Some of the wireless applications provided by WASPs include providing access to customer relationship management, mobile workforce management, and *transaction enablement* (the service that allows mobile users to check item pricing and availability, and engage in mobile commerce).

The hosting market has undergone severe consolidation over the past few years, with the major companies now accounting for the bulk of the business. Over the years, several larger companies have entered the space, including hardware manufacturers such as Dell, Hewlett-Packard, and IBM, and major carriers such as ATT, MCI, and also Sprint, which announced in mid-2003 that it would exit the business again. The large companies are wielding their strong brands and established customer bases to expand their share of this market. Their goal, of course, is to leverage their cache with the much more profitable corporate clients. But as barriers to entry are not that high in this segment—setting up a hosting operation requires only a few servers and fast connections to the Internet backbone—this market is seeing a constant ebb and flow of smaller companies, engaged in a cut-throat competition for the less profitable small business and personal Web site market. Critical success factors for large hosting companies include the ability to form profitable relations with those organizations that need to host content, and the ability to compete with small contenders on price. WASPs must forge alliances with application developers whose packages are likely to emerge as the winners in a still-evolving marketplace.

Technology Enablers

The second orbit around the end-user at the center of the Wireless Value Web contains the segment we call technology enablers. Technology enablers, as illustrated in Figure 6.5, provide the products and services that enable, facilitate, or enhance customers accessing content via wireless networks. In contrast to the hardware and software providers mentioned within the first orbit of the web, the companies covered in this segment contribute offerings to the network that are not directly handled by the consumer, but instead operate out of sight and out of touch.

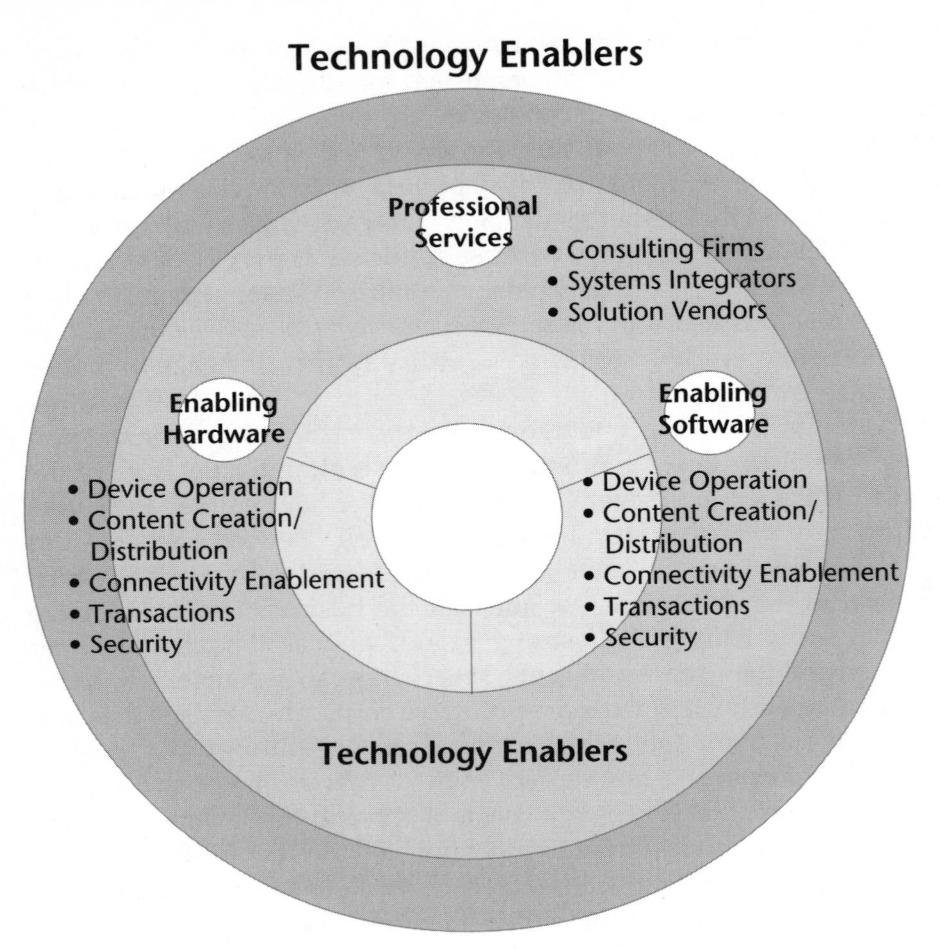

Source: CGE&Y, J. Teutloff

Figure 6.5 Technology enablers provide the products and services that enable, facilitate, or enhance customers' accessing content via wireless networks.

Within the technology enablers segment, we find the following categories of companies: professional services providers, enabling hardware providers, and enabling software vendors.

Enabling Hardware

Let's start with the providers of *enabling hardware*, hardware that makes wireless work behind the scenes. We previously presented three Value Web segments in the first layer that wraps around the end-user. These three segments were user devices, content and connectivity—which provides us with a convenient way to structure the discussion of the physical devices used by them. Each of these three customer-facing segments operates on hardware that is provided by companies within the enabling hardware category. The primary physical technologies that you find when removing the outer case of your cell phone, PDA, or other increasingly sophisticated end-user gadget include various types of microchips (subscriber identity modules (SIM cards), microprocessors, analog-to-digital and digital-to-analog conversion chips, digital signal processors, memory chips), radio-frequency amplifiers, microphones, speakers, LCD (liquid crystal display) or plasma displays, antennas, batteries, keyboards, input/output ports, and other electronic components.

Enabling hardware for the content providers includes the computer systems and the various input devices the written, visual, or audible content is created on. A full discussion of the many devices—most of them the same as used in the wired world, including graphics workstations, digital cameras, scanners, printers, musical keyboards, and so forth—exceeds the scope of this book. Similarly, enabling hardware available to content aggregators and application developers is omnifarious. Suffice it to say that there is a seemingly endless supply of devices on the market to assist content providers in their core activities of creating, presenting, and distributing content, with new or improved technologies appearing almost every other day.

Enabling hardware for connectivity providers, such as network operators and wireless service providers, entails wireless LAN hardware, network interface cards, wireless routers, switches, hubs, transceivers, firewalls, access points, mobile switching centers, base stations (towers), base station controllers, and a slew of other, transmission-related components. Hosting companies rely on HTTP (Hypertext Transport Protocol), SMTP (Simple Mail Transport Protocol), and NNTP (Network News Transport Protocol) servers, mobile application servers, transaction, video and chip servers, and other devices, again too many to discuss in detail.

Last, we need to briefly mention the groups of companies that provide the behind-the-scenes hardware that enable wireless transactions and security. Examples of enabling devices surrounding security include, for example, those companies that manufacture and distribute biometric identification technologies that verify the user's identity via fingerprints, voice prints, or retinal images before the wireless device can be activated. Such devices are still very expensive and just recently begun to enter the mass market. Although the more complex identification systems, such as retinal scanners, are currently confined to high-security environments, there are now PDAs available that are activated by your fingerprint.

Enabling Software

The next category within the technology enablers segment comprises those companies that provide *enabling software*. Enabling software is a catchall term for a long list of software that ranges all the way from programming languages to operating systems, microbrowsers, development platforms, security systems, middleware, and application programming interfaces. In essence, this type of software is "the brains" behind the functions that add value to the Wireless Value Web. Enabling software runs the devices we have talked about; allows providers of information to create, format, and distribute their offerings; securely manages data and their transfer between network nodes; and provides the back-end mechanisms to complete m-Commerce transactions.

The most critical enabling software for user devices includes operating systems, browsers, and other programs required to run the device. Remember that we do not include end-user applications in this category. The reason, as you will recall, is that end-users directly interact with these applications, which places them into the first orbit around the customer. End-user applications are part of the application developers category of the content segment. Enabling software is running in the background, mostly invisible to the customer handling the wireless device. Critical success factors for these companies are those that make the company's product become a standard. Whether operating system or browser, once the software has found its way into a large base of handsets or PDAs, its platform status provides long-term viability as application developers build their programs upon the standard.

Content providers equally rely on enabling software; authors use these programs when designing images, text, or sound. Aggregators use the software to build and maintain their portals, commercial Web sites, product catalogs, customer profiles, personalization, user tracking systems, and so on. Application developers use various software tools such as programming languages, development platforms, debuggers, and so on, to assist in the coding of their educational, entertainment, communication, or informational or commerce-related programs. All require appropriate middleware to deliver content to an ever-increasing array of wireless devices.

Connectivity providers require enabling software for efficient network monitoring and hosting management in a secure environment. Specifically, network operators and wireless service providers are concerned about reliable connectivity and secure voice and data transmissions. Hosting providers, managing the content to be delivered to the user device, rely on middleware for effective device detection, optimization of content delivery by type of device, compression technology, and transaction mechanisms.

Other companies also located in this group provide access to corporate data, that is, the enabling software by which existing content—including applications and data—is ported, say, from a company's Web site to a wireless device.

Especially when it comes to completing mobile business transactions, enabling software is critical to ensure secure data transmission, rapid and accurate billing, payment verification and settlements, and other processes that support wireless communications and commerce while integrating with the company's back-end systems. For example, providers of billing software enable carriers, ASPs (application service providers), and ISPs to charge their customers for voice and data services used. Other applications track wireless services usage, and allow for customer self-service via a carrier's Web site. Yet other billing-related functionality ensures that revenue is shared between a carrier and a content provider according to agreements reached by both parties.

Providers of security-focused enabling software are offering virtual private networks, authentication software for mobile employees, public key infrastructure security, digital certificates, and other applications that act as safety measures in the Wireless Value Web.

Professional Services

A third category within the technology enablers segment includes *professional services*. This category includes consulting companies, system integrators, and solution vendors that provide professional services that enable or enhance the wireless experience.

Cap Gemini Ernst & Young, one of the world's largest consulting firms with 60,000 employees in over 30 countries, assists clients all the way from formulating a mobile technology strategy to designing, developing, and deploying mobile and wireless technologies. CGE&Y has built full-scale pilots as part of the development of advanced business portals with a focus on finance-, travel-, retail-, and employee-centered systems. In addition, the firm has created and implemented a range of key corporate mobile applications, including mobile marketplaces, telecommunications solutions, supply chain applications, telemetry, fleet management, job dispatching, and all aspects of CRM solutions. Other applications that CGE&Y has developed include those for use by the public in their own homes, such as mobile banking and brokering, payment, shopping, and advertising solutions.

In addition to custom development and package implementations, CGE&Y is providing systems integration services. According to the Forrester Research, approximately 45 percent of the costs in e-commerce projects are related to systems integration. We've all heard the mantra: "Integrate your processes with what you do on the Web," forcing the integration between Web and legacy systems. The key to success lies in the integration of business processes, ERP, CRM, and supply chain applications, and mobility, an area where CGE&Y has established a leadership position, building on years of application integration expertise and thousands of completed projects.

Aside from enterprise consultancies and systems integrators, solution vendors themselves are frequently building consulting services around mobility—most frequently directly related to the company's offering. For example, BEA Systems, IBM, Oracle, PeopleSoft, and SAP have developed professional services groups within their organizations that assist the company's customers with making the right choice regarding technologies.

Environmental Enablers

The final layer of the onion, so to speak, contains a segment we like to call environmental enablers, as illustrated in Figure 6.6. The segment includes trade associations, regulators, standards-setting entities, research firms, and providers of financial services. Environmental enablers provide the products, services, and information that facilitate the entire workings of the industry. They set the foundation upon which economic activity can grow. To use a familiar analogy, similar to a chemical reaction, enablers also act as *catalysts* that can launch and accelerate the business activities of all other Wireless Value Web participants.

Trade Associations

Trade associations are organizations of companies that are aligned by virtue of their common interests. Just as any other industry's members organize into trade associations, companies in the wireless industry gather for the purposes of information exchange, leading practices sharing, quality standards definition, or to promote their offerings to members and interested parties outside of the association. In addition, trade associations frequently represent their members in communications with standards associations and legislative regulators to promote the association's members' interests. Trade associations, frequently aligned along specific product or service categories, organize trade shows, commission and publish research studies, certify products, and engage in consumer education and information.

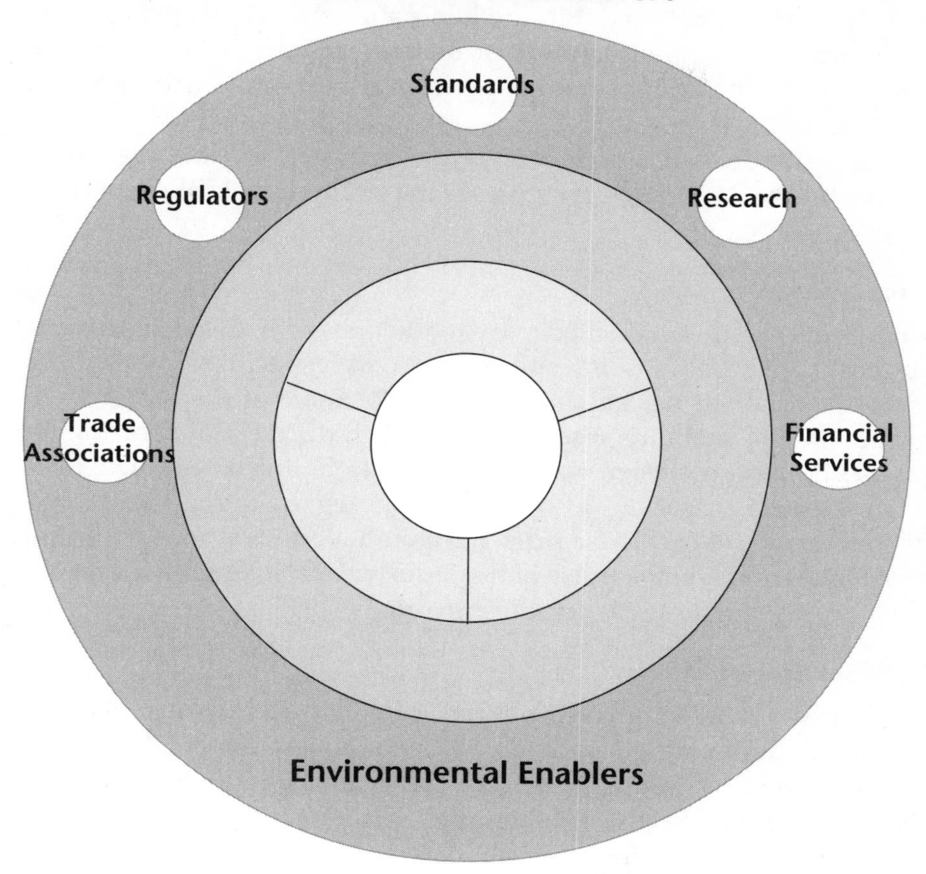

Source: CGE&Y, J. Teutloff

Figure 6.6 Environmental enablers include trade associations, standards associations, regulators, research firms, and financial services providers.

Standards Associations

Standards associations, usually independent nonprofit entities, are primarily concerned with the definition and development of technical standards upon which wireless systems and applications are built. Standards associations further engage in promoting their approved standards to enable industry-wide adoption.

Regulators

Regulatory bodies at state, federal, and international levels are policy makers in the governmental structure that oversee various areas within the wireless industry, including wireless spectrum allocation, security, privacy, and other areas.

For a list of trade associations, standards organizations, and regulators that illustrates the wide range of entities in the environmental enablers segment, see Appendix A.

Research Firms

Research firms mostly conduct primary market research, develop insights, and publish their findings in reports that cover specific topics within the industry, or about the industry as a whole. Some of the well-known companies in the mobility research space include the Cahners In-Stat Group, Dataquest, Forrester Research, Frost & Sullivan, Gartner, IDC, Herschel Shosteck Associates, Ovum, The Strategis Group, The Yankee Group, and many others. These firms have built a business model around keeping their fingers on the pulse of the industry and selling their diagnosis to whomever purchases a membership or pays on a per report basis.

Financial Services Providers

Another category of companies offering research and analysis similar to that of the pure-play research outfits includes investment firms that cover the mobile technologies market. Bear Stearns, CS First Boston, Durlacher, Goldman Sachs, Lehman Brothers, Merrill Lynch, Morgan Stanley Dean Witter, Salomon Smith Barney, and others prepare such reports for their customers or any other interested party willing to pay for these documents. Their main function, however, is that of a provider of access to capital sources. Financial services providers, including venture capitalists, investment banks, commercial banks, fund managers, governments, and private investors are types of organizations that provide financial advice and capital to organizations within the Wireless Value Web.

Before we close this chapter, here is a look at the comprehensive picture. Figure 6.7 summarizes the Wireless Value Web as we have presented it on the previous pages.

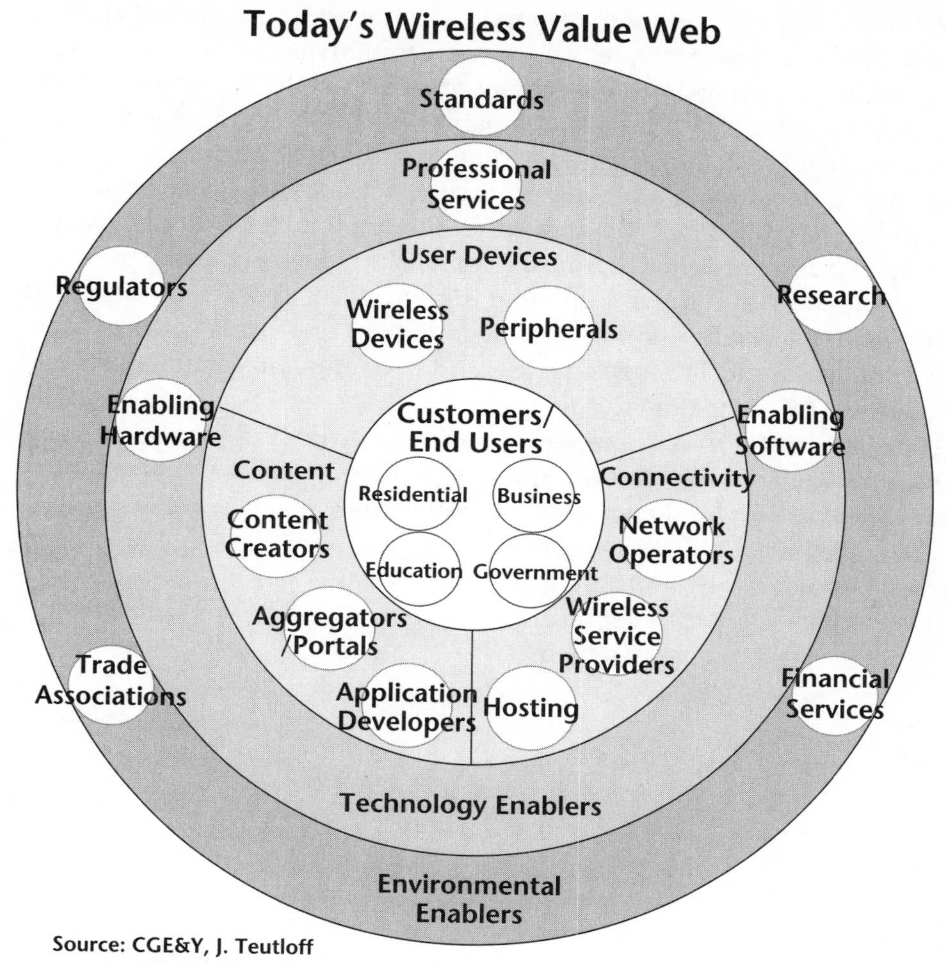

Figure 6.7 Today's Wireless Value Web.

Summary

Let's summarize the key take-aways of this chapter. We started by introducing the Wireless Value Web and its origin. The key point was that you can construct Value Webs not only for specific companies and their business environments, but also for entire industries. The Wireless Value Web we presented in this chapter is frequently used as an analytic tool created by CGE&Y's mobile commerce group to serve as a stepping stone in creating mobile/wireless strategies and tactical initiatives for the firm's clientele. The tool's primary value lies in its capability to illustrate in a concise, organized matter the types of companies currently vying for the customer's attention. A thorough understanding of the wireless universe is a necessity to developing your own course of action. However, it is important to keep in mind that the wireless industry picture we have drawn in this chapter is a snapshot in time as the market space is rapidly evolving with companies coming and going at a rapid pace. Yet the taxonomy we introduced, especially the industry segments of end-users, devices, content, connectivity, technology enablers and environmental enablers, should hold steady for some time to come.

The next chapter introduces you to various mobile and wireless applications and case studies that may spark your own ideas about how mobile and/or wireless technologies may add value to your organization.

CHAPTER 7

The Three Functional Domains

Michael Welin-Berger

Introduction

We previously provided you with a platform that illustrates the various actors within the Wireless Value Web and how they interact with each other. Understanding the key players is a critical requirement for designing functionality that corresponds to a company's business and technology requirements. Having established this platform, we are now ready to choose what mobile functionality will provide the most value to an organization and its business processes. However, mobile technology is complex in that it can be applied to almost any business process, in any industry. Thus, we need to create a structure for our discussion. To make it easier to present specific examples for mobile applications, we have divided mobile functionality into three major domains: communication, information, and commerce.

NOTE In this discussion, a *domain* is no more than a broad category that can be easily divided into a subset or market segment of a business.

- **Communication.** This domain encompasses all communication between individuals and/or organizations. The most common applications are voice, voice messaging, e-mail, SMS, alerts, and instant messaging (IM).
- **Information.** This domain addresses access to different information sources. The Internet is the best example of where you find information for private use and for business process support.
- **Commerce.** This domain is a combination of communication and information with the purpose of buying, selling, or improving services. Commerce does not necessarily mean that you pay for the service with a mobile device.

Within each domain, there is a need to separate how the usage will support consumers on the one hand and enterprise processes on the other. For this reason, we have divided the domains into consumer applications and enterprise processes. Entertainment can constitute a fourth domain, but because we are focusing on gaining business value, we have chosen to address entertainment separately in Chapter 9.

Communication

Communication is essential to everything we do, and face-to-face dialogue is definitely the most efficient communication tool we have. Communication is most efficient when we combine speech, facial expressions, and gestures. For instance, we can attach emotions to words through facial expressions or describe the shapes of objects with the help of our hands. Think about this the next time someone describes an event to you—regardless of the situation you are in, you will see that people communicate using all these elements. In order to allow us to communicate as effectively as possible, mobile devices have to combine text, voice, pictures, and ultimately video.

Choosing which tool we want to use when we communicate has become a daily task. Some topics demand that you speak to another person directly (by using a phone), whereas other topics are best communicated via text (e-mail). A text message would for instance suffice when ordering a spare part you already know that you need. Conversely, in-depth consultation with a colleague might be necessary when deciding exactly which spare part needs to be replaced.

New communication opportunities are available to us today—videophones are one example of this—but they can only be used on a fixed line. We will, however, soon be able to use mobile videophones with which it will be possible to see the person you are speaking to and show him or her objects as you are describing them. A small built-in camera in the device will enable this. We now need to understand how these new mobile functionalities can support communication. Let's begin by describing in what situations they can be useful and then look at the best way to use them.

Consumer Applications

Today, you have to check several systems separately to see whether you have messages. In other words, you use one device to check e-mail and another to check your voice messages. When you want to check your e-mail while on a business trip, you are dependent upon Internet cafés or a slow connection via your cell phone. This is one of the tangible drawbacks to the lack of connection between your e-mail and voice messages.

The use of cell phones has increased dramatically—just about everyone has one and penetration has been faster than anyone could have expected only a couple of years ago. However, almost all communication today is voice-based. Adults use cell phones to communicate with their children and friends, and spontaneous business meetings are becoming increasingly common because it is easier to get hold of people. The elderly use cell phones not only to communicate with friends and family, but also as a precautionary device in case something should happen when they are taking a walk, driving a car, and so on. We now want to extend this voice communication and combine it with text and graphics in one device.

In the future, you will be able to choose the most suitable device for communication at any given time. The only factor that will dictate your choice is the situation you find yourself in—it will be up to you to decide how you want to have the information presented to you. You will, in short, be able to not only get to your voicemail, but also access e-mail, faxes, and other messages whenever you want, from wherever you are. All you will have to do is to choose between having new messages presented to you in text, voice, or video format. This is called *unified messaging*. This topic is covered in more detail in Chapter 8.

Here are some concrete examples of the choices you will be able to make (also illustrated in Figure 7.1):

- Listening to a message takes a long time. A relatively short message may take as long as 1–2 minutes to be read to you because, in addition to the message itself, you will have to listen to who sent it, what priority status it has, when it was sent, and so on. However, when you have time on your hands (when sitting in your car on your way to the office) this is a perfect way of getting updated.

- Reading a relatively short message on a screen only takes about 5–10 seconds. We have trained ourselves to quickly look through our e-mail inbox to see which messages need to be prioritized (something that can't be done when listening to messages). Therefore, reading your messages will be the obvious choice when you have access to a screen.
- Several devices will be available to you when you want to access information. You will simply choose between devices, depending on the situation you are in at any given point in time.

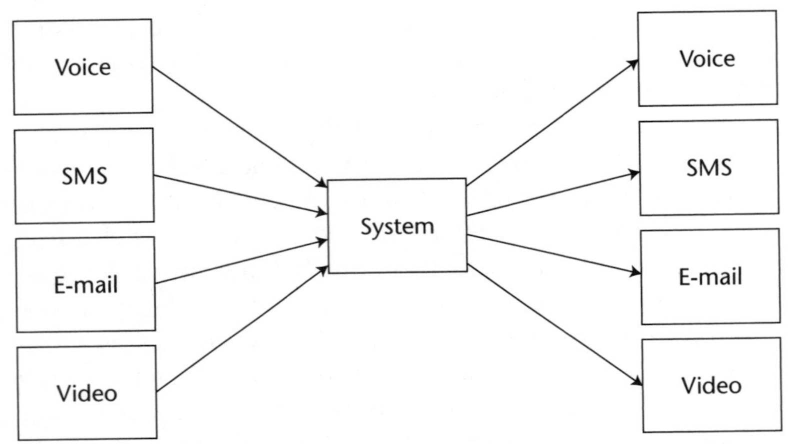

Figure 7.1 Providing and receiving information through different devices

Enterprise Processes

Enterprise processes have experienced a revolution—communication between individuals (or functionalities) is much quicker today than it was only 10 years ago. The use of cell phones, e-mail, and even automated messaging (alerts) from IT systems has virtually exploded. However, we are still using PCs and phones as our main devices for exchanging messages. This means that a mobile employee often places calls using a cell phone when an e-mail would actually have been more efficient.

Now, with the use of mobile functionality, we will take one more step toward streamlining our processes and cutting lead times. One of the most important changes that will take place is that we will reduce the amount of paperwork that the mobile worker has to deal with—in much the same way we have done at the office—because most work orders and instructions will be performed via mobile devices (this will be addressed in more detail later in this chapter). The new mobile functionalities will also provide mobile workers, back-office staff, and management with information of higher quality because everybody will feed the back-office system directly.

The major advantage of sending a text message from a mobile device rather than making a phone call is speed—your main concern is getting a quick response. Text messages can be sent in 15 seconds and they don't require that the recipient be available. Phone calls, on the other hand, take at the very least 3 minutes to make. Similarly, sending text messages from the new "mixed" devices—such as PDAs or smartphones—will be faster and more efficient than sending traditional e-mails (you will receive replies within hours instead of within days). It is important to point out that this requires a common culture in which people check their messages frequently. They will also be expected to have activated an autoreply feature when they are not connected to the system for more than, say, one day.

Another advantage is that sales reps or field technicians will be allowed to work without interruption while still being available as soon as they take a break. This assumes, of course, that they are performing a task that allows them to check for messages on a regular basis. Today, most mobile workers turn their cell phones off when they are working. When they turn them on again they have to listen to a number of long voice messages, most of which would have been more easily dealt with if they had been in e-mail format.

Information

"Always connected" will be one of the most compelling features introduced through new mobile services. There will be no need to endure a 3-minute startup phase in order to access the Internet. What's more, you will no longer be restricted to using your PC when you want to be connected. Instant access to the Web will make people use their devices several times a day, for both private and business purposes.

Consumer Applications

When the "e-boom" took place in the late 1990s, the perception was that practically everything would be purchased on the Internet. But that didn't happen. Instead, people began to gather information on the Web—they rarely actually bought anything. Let's look at some examples of this behavior. People don't normally buy cars on the Web, but many people use it to find any relevant information on a car they want to buy. So, they still buy a car through a car dealer but they have more or less made up their minds beforehand based on information they have gathered on the Web. All that's left is a test drive and a discussion about the car they want to trade in.

People also frequently turn to the Web for information services, such as finding a telephone number, getting driving instructions, finding which cinema is playing the movie they want to see, and the like. These services do, however, have limitations. Driving instructions, for instance, have to be printed out if you want to have access to them in your car. Similarly, the telephone directory is only available when you are connected to the Internet with a PC.

These services have one characteristic in common: they would be much more useful if they were available to you at all times—not only when you are sitting in front of your PC. This is what tomorrow brings in the form of new mobile services. You will have access to the telephone directory when you are walking down the street and want to call your client because you are late for an appointment. The same goes for driving instructions. Whereas today's solutions force you to plan your day in advance, tomorrow's mobile services will make information available when and where you need it.

That very concept, "getting information when and where we need it," could be the slogan for the mobile world because it summarizes what will be delivered. This mobile functionality will be used in combination with services from the PC at home or at the office. It will also be exciting to see what new information services will be available when *location awareness* (knowing where we—our devices—are located at any given point in time) is combined with our information needs.

Enterprise Processes

The description of the enterprise process becomes increasingly important when information is becoming available in all situations. You need to define the task in which information could improve your performance. Process management was hot during the late 1980s, more or less forgotten during the Internet hype in the 1990s, and increasingly important with mobile applications.

Information Available to Your Customers

Company brands will become increasingly important with the extended use of the Internet. Customers will be presented with a simple way of contacting a company if they want to know more about certain products or services. Today, practically all advertising directed at potential or existing customers is presented along with a Web address for more information, regardless of whether it is distributed via direct marketing or standard advertising. Tomorrow, companies will need to target potential customers who are on the move. A person might see a product in a shop and want to know more about it before buying it. Any information that can support mobile customers (when they are not close to a PC) as they are in the process of buying a product will be instantly accessible from their mobile device.

Information Available to Your Employees

Most companies have content-rich intranet solutions, which normally are a goldmine for employees who need information about their company and/or its products or services. This type of information will be available to employees at all times. As a result, when employees meet a potential client, they will be able to show and describe products and services in a more informed manner.

Here are two examples that further illustrate the benefits of having information available while on the go. When a technician is working at a client's premises, he or she will be able to answer questions concerning the number of available products, quantities, or prices right away. Corporate information will receive high priority; top management will be adamant about communicating what is happening inside the company to employees who are working outside the corporate buildings. This has traditionally been hard to achieve.

Commerce

Commerce was expected to become the major revenue generator on the Web. It has, however, largely failed; too much focus was placed on the payment part of the purchasing process instead of on supporting the process as a whole. Hence, the concept "buy, sell, or improve services." The intention is to move from a pure purchase perspective to a perspective that encompasses the entire process; that is, obtaining information, purchasing a product or service, and delivering the chosen product or service to the customer. When using the Web—indeed, our new mobile devices—we will need to focus on the entire process, much in the same way that many corporations currently use the Web to communicate with their potential and existing clients. For example, it will be very common to provide full-service support via the mobile device when any kind of malfunction occurs. Similarly, you may find yourself researching a movie you want to see, selecting show times, finding the nearest location, and engaging in the transaction—all from your mobile device.

Consumer Applications

Improved commercial processes will result from a combination of communication and information with the purpose of buying, selling, or improving products and services. We sometimes forget the benefits of separately enhancing different steps of the commercial process by focusing too much on the transaction being performed through mobile devices. We can simplify the process by dividing it into six steps:

- Identifying the need for a service or product
- Researching the service or product in relation to the need—price, specifications, and availability
- Purchasing the product or service
- Receiving the item
- Obtaining customer support
- Evaluating the product or service

The step that involves a transaction is the third step, the actual purchase. The ability to order products and services through a mobile device increases the probability that the purchase will really take place. We often say to ourselves: "I'll do this when I get home." But by that time we've often forgotten what we were going to do or have simply changed our mind. Enabling the customer to be spontaneous, to order something as soon as an idea springs to mind, will increase revenues for products and services that are appropriate for purchase over a mobile device. Although you might not purchase your next car from a cell phone, you might engage in an impulse purchase of items that require less involvement, require less information gathering and evaluation, and are relatively low in cost.

Let's look at an example. Using the Internet to send flowers to your loved ones is becoming increasingly common, but you usually have to do this from your office or from a hotel you're staying at. Now, you can order the flower while waiting in an airport lounge when it dawns on you that your tenth business trip in a month was not appreciated back home. Books and music CDs are currently being bought extensively on the Web. These purchases are normally of things you know by name, which will make them easy to order via a relatively small screen. Booking a time at the Department of Motor Vehicles to inspect your car will also be a relatively simple transaction—and it will be made even simpler if you have your calendar available to find open slots that fit your schedule.

Integration with enterprise systems, the enabler for these activities, will play a key role in extending functionalities to mobile devices. We will use the same functionalities currently being used on the Web and simply extend them to mobile devices in the early stages. Many functions will be supported through mobile devices such as phones and PDAs without the need of complex integration; the key will be to establish which functionalities will provide the best support for the mobile user.

Enterprise Processes

When we look at the evolution for mobile commerce, it has failed in the sense of being a sales channel. However, mobile applications can provide much more efficient processes when employees can use them for collecting information (such as the use of bar codes). Information quality is another area in which mobile applications support improvements.

Enhanced Processes with Your Customers

The examples described previously illustrate benefits for the client, and they can be summarized as "better service." The goal is to make it easier to buy a product or service and to be able to do so anywhere and anytime. We will also see increased revenues generated by spontaneous purchases that otherwise might not have been made.

Enhanced Processes with Your Employees

This is an extensive area that we cover in depth in the sections "Field Service" and "Sales Force" in Chapter 10. Here is a brief overview.

Two things in particular will be improved as a direct result of providing the employee with mobile functionality. First, there will be improved information available to the client during meetings and better-prepared employees when they meet clients. Knowing what is available in stock and which items should be promoted due to imminent expiration will lead to increased efficiency and profitability. This will also give the customer a feeling of improved service; an example is the ability to receive an answer concerning when deliveries can be made during an actual meeting.

Second, the number of administrative tasks will be reduced—both for the mobile employee and for back-office staff. When orders are registered directly in a client meeting, no papers have to be handled afterward. The administrative process will, in other words, be largely completed during the meeting. This will lead to huge efficiency gains within the company.

Normally, there is one back-office administrator for every 5–10 mobile employees. As a result of mobile solutions, the support staff can be reduced by half. For mobile employees, the stack of Post-it notes will disappear, and they will gain control over the entire transaction. This has proven to be the easiest area when building a business case and calculating an ROI (return on investment) for a particular initiative. Cost savings due to a reduction in administrative staff is substantially easier to calculate than the expected top-line revenue a client might gain from deploying mobile functionality.

When building applications, the most common improvements result from extended functionalities that can be used when meeting a client as well as reduced back-office costs. Reduced costs provide hard figures to justify the investment, whereas qualitative benefits have to be calculated by way of rough estimates.

Summary

By dividing the structure of this chapter into communication, information, and commerce, we have attempted to provide you with the different possibilities available, depending on your situation at any given time. This division is important because mobility is a very wide area, which at times will make it difficult to identify the best solution, as it can be applied to any process.

Communication, information, and commerce will overlap when building any kind of functionality. However, it is important to first identify the core need that can be met with mobility. Second, it is necessary to identify relevant functionalities and the channels that can be used to support them.

Chapter 8 looks at how mobile devices use the combination of text and voice when communicating. This is an area where we have seen great improvements recently.

CHAPTER

8

Communication

Michael Welin-Berger

Consumer Applications

New communication opportunities are available to us today. What we need to do now is to learn how to choose the best alternative when we want to communicate.

Choosing what tool we want to use when we need to communicate with someone is easy when we restrict ourselves to the tools we are familiar with today, such as writing a letter, sending an e-mail, or calling someone (from a cell phone). However, the options are increasing with instant messaging (such as SMS) and new services, such as sending a video sequence to a cell phone.

We will describe both existing services and services that will be available shortly. The rollout of these services is contingent upon the infrastructure in different geographies as well as the expected acceptance of these services.

Voice, E-mail, Paging, SMS/EMS, and Video

The section provides an overview of the available options, looking not only at the different services that exist on the market, but also at how people use different the interfaces, such as voice, text, and video.

Voice

The use of voice communication has gone from fixed-line to mobile communication thanks to the cell phone. We now frequently leave voice messages on different kinds of answering services connected to fixed or mobile subscriptions.

Voice messaging was the first step in mobile communication, and it has increased our availability in both our private and professional lives. The second step was to use voice to choose between alternatives in an application. One example of this is telephone-banking services, which are already being used quite frequently. When compared to numerical input ("choose four for your savings balance"), which is the most common form of telephone banking service, using your voice to make choices has several advantages. Current systems have relatively simple functionalities, and you can only choose from a small group of commands. However, more advanced functionalities will eventually emerge.

Voice communication is not, however, the optimal solution in all cases. Obtaining information through voice communication is often far slower than if the information is presented to you on a screen. In specific situations—such as when you are driving your car and it would be dangerous to look at a screen—the best communication alternative is voice. Driving is one relevant situation. Another is when you don't have your smartphone or PDA available; you can still use these services from any telephone on earth.

Let's take a look at two specific voice applications: voice-to-text and text-to-voice. Text-to-voice applications are easy to create and there are several players on the market. The voice quality is getting better—it doesn't sound as if it is coming from a can anymore, which used to be the case. One application that is being used increasingly is having e-mail read over the phone. Simply by activating the answering service and providing your access code, you can listen to all your e-mails. All the functions that are available in the traditional PC interface are also available in the voice interface (functions such as reply, delete, and move to a specific folder).

Voice-to-text, on the other hand, is a much more complex application that poses a number of difficulties. Getting a computer to actually understand a sentence and then convert it into text is a very complex process. This demands that the application understand language in all its forms—including dialects, slang, and incorrect language use. This is, naturally, relatively difficult to implement when designing an application. Another issue that complicates voice-to-text communication is the processor power that's needed. Today, only powerful PCs can process this kind of application with enough speed. Mobile devices currently don't have this capability, but hosting the applications intended for mobile use in a central server will solve the problem.

The number of voice-to-text applications will grow, but they will be customized for such areas as hospital administration (for example, registering doctors' notes and the like). Voice-to-text will be a widely used feature when the technology has matured, but there are two hurdles that need to be overcome:

- Translation applications have to become better at voice interpretation.
- The performance of mobile devices has to be improved in order for voice interpretation services to be transferred from a server to a mobile device.

E-mail and Unified Messaging

We all use e-mail for many purposes—both privately and for business. We use e-mail to transfer information, to send requests, or simply to say hello to someone. E-mail is now being extended into *unified messaging,* which mixes voice and text presentations. This applies to both the input and the reception of information.

Two services are required for unified messaging: voice recognition, which transforms your voice messages into text messages, and a synthetic voice service, which transforms your text messages into voice messages. All messages are then stored in the same server, and it doesn't matter how the message is delivered to the service (via voice or text). Furthermore, you will always have the choice of either reading or listening to your messages, depending on the situation you are in.

The need to prioritize different messages will increase with the growing number of messages you send and receive. Today, we normally prioritize messages based on their format. For example, voice messages often get higher priority than e-mails—they are simply deemed more urgent. We are all aware of this and that is why we often call someone as soon as we have sent an important e-mail—we want to make sure the e-mail has been received. The next step is to learn how to use functions such as "high priority" to ensure that a message is read or listened to as quickly as possible.

In the future, communication will be simpler, and you will have to carry only one device to send or receive messages in different formats (for example, voice and text). Mobile devices will be a combination of PDAs and cell phones, capable of either reading messages to you or displaying them on a screen. Thus, the most important difference from today's standard is that we will get the information in the format we want it, when we want it. Instead of waiting to be connected so that you can read your e-mail, you will be online all the time!

Let's look at some concrete scenarios. Imagine that you are waiting at an airport and you have 20 minutes to kill before your plane takes off. You suddenly remember that your son has to buy his mother a present for Mother's Day. You are faced with two options: you can either send your son a message directly or you can make a mental note of it (which you will probably forget). Needless to say, the former option is preferred. Similarly, if you get a toothache, you can instantly send a message to your dentist saying that you need to book an appointment. Being mobile is very much about doing things when they spring to mind instead of having to make a note of them.

Constantly being available through mobile devices is something we will have to learn to cope with. We are learning how to live with cell phones. Although most people can handle the new technology, others experience stress because they become slaves to the new technology instead of mastering it. If we learn how to use technology as a tool rather than allowing it to dictate what we do minute by minute, we will have much more time on our hands.

Pagers and BlackBerrys

Paging had its "days of glory" in the 1980s and 1990s, depending on which part of the world we are discussing. Pagers can receive short text messages, usually consisting of a telephone number that someone wants you to call, status codes, other short messages. Their main advantage is good geographical coverage, while their main disadvantage is that they only provide one-way communication—you can't confirm that you have received a message. The pager eventually received serious competition with the introduction of two new devices: the cell phone in conjunction with SMS, and the *BlackBerry*. The BlackBerry is an extended pager that can both send and receive e-mails. It is built on the same communication technology as the pager. The BlackBerry originated in the United States but has now spread to other countries. The BlackBerry from RIM is simply a PDA that can receive and send e-mails. Its size, long-lasting batteries, and user-friendliness make it an interesting alternative for various user groups. The latest versions include different networks to enable communication outside the United States.

SMS/EMS/MMS

SMS (Short Message Service) is a simple way of sending short information that can be viewed on any device, mostly on cell phones. The major advantage is that you can send a message from your phone (you don't need a PDA or a computer), and recipients can read it on their phones. It is very easy to read the information. A maximum of 160 characters is allowed, but that limit is rarely exceeded because it is relatively difficult to write the message using the phone's keyboard. For this reason, many people use computers to send an SMS—which is then read on the recipient's phone. Young people all around the world have taken different versions of SMS to heart and use it frequently. SMS is also an alternative to paging services and can be used to provide such services as sending reminders, receiving stock and currency quotes, making airline schedules, and accessing bank account information.

EMS (Extended Message System) is SMS with added functionality that allows you to include pixelated pictures and animations, sound effects, ring signals, and formatted text in the message.

MMS (Multimedia Messaging Service) is EMS with added functionality that allows you to include speech and audio. A multimedia message can, for example, be a photo or picture postcard annotated with text and/or an audio clip, a synchronized audio playback, a text, a picture, or a video clip. It can also simply be a drawing with added text.

Video

You can currently view video clips from several different mobile devices. The main advantage of this service is that it allows you to videoconference with other people using your mobile device. The adoption of this service will probably take some time since we have seen a slow adoption rate of videoconferencing on the Internet, despite the fact that it has been available for several years. There are no major technical difficulties except for the bandwidth required, which is becoming available in 2003 and 2004. The first area of use will probably be to download news and weather forecasts to your mobile device, giving you the opportunity to watch the news while you are waiting for someone.

Office Applications and Internet/Intranet Access

Mobile devices can also be used for office applications and Internet/intranet access. This will differ only slightly from the way you use a laptop—the small screen and the lack of a traditional keyboard are just about the only restrictions to what you can do with your mobile device. The difficulty in providing input probably means that mobile devices will predominantly be used for reading and searching purposes. Input will instead be carried out on traditional PCs. Access to Web applications on the intranet or Internet does not pose the same problem because Web applications are designed to reduce text input, through, for example, the use of drop-down lists, where you simply choose one option.

Office Applications

You will not use your mobile device to create traditional office documents, but you will have access to all the files you need, when you need them. Small changes can, of course, be made but the screen size and lack of a real keyboard will make performing substantial changes both tedious and time-consuming. Therefore, the combination of the traditional office environment with a desktop computer or a laptop is required. You will be able to access files either by connecting and then downloading information or by synchronizing the mobile device with your PC. *Synchronizing* simply means that data changes, either on the device or on the server, are exchanged, thereby providing the same information in both places. You will most likely have your frequently used files synchronized with the PC and stored in the device, whereas the files you normally do not use can be downloaded when necessary. All office suites will contain formats for word processing, spreadsheets, and presentations that can be viewed and altered on all mobile devices.

The data formats used by word processors and spreadsheets to store information in mobile devices are usually reduced if you compare them with traditional PC formats (such as Word or Excel). The data formats used in the mobile devices will normally be simplified versions that won't have all the formatting features offered by a PC.

Internet/Intranet Access

When the device is connected, there will be no constraints regarding which services you can use. All applications available on the Internet or the corporate intranet can be accessed using standard security and performance functionalities. The challenge for the Web sites will be to make the functionality as simple to use on a mobile device as it is on a connected PC—a difficult task considering the small screens on mobile devices. Everything you normally do on your PC at home will be possible on your mobile device—for example, all Web sites will have the same address as on your PC. This will make things simple from the word go. There will be no need to learn a new structure for finding information, and you will be able to apply all your earlier knowledge of the Web. This means that finding a movie theater or a friend's telephone number becomes as simple as surfing the Web via your PC. The difference is that you will have this capability no matter where you are—whether you're in another country or in the countryside. Your bank will provide you with Web access, so you will be able to carry out financial transactions from anywhere in the world. You'll purchase books online and purchase and listen to music online. You will also be able to make all your travel arrangements from your device, changing travel times or hotels when you're on the move.

For larger corporations, the corporate intranet is a rich source of information, so they will want employees to check in regularly to get the latest corporate information.

Enterprise Processes

When looking at different functionalities for consumer applications, you'll naturally see the same usage in the enterprise processes. However, there are specific gains to be made by enhancing processes within an organization. This section identifies those gains.

Voice, E-mail, Paging, SMS/EMS, Video

Cell phones are the devices of choice when communicating with employees who are not at their desks. This behavior will continue, but new channels will support it. Many people think that e-mail is the nail in mobile usage's coffin. But that isn't the case. In fact, there is no single killer application in the mobile enhancement of enterprise processes—enterprise efficiency will be optimized as a result of combining different communication channels.

Let's take a closer look at some of the technology surrounding messaging so we can get away from the view of devices as mere message repositories. Messages will always reside in a central server. You can then synchronize and download messages to your mobile device or use a browser to view them. The key element is that you will use different devices all the time—probably a PC when you are at your desk, a mid-sized PDA when you have just left your desk, and a smartphone when you are traveling. The increased speed in the wireless network will enable you to handle your messages almost as fast as when you are connected to a fixed line—the only restrictions will be the screen size and the keyboard. Thus, you will learn to prioritize devices according to the situation you are in and which device is at hand (you will not use a palm-sized device with a touch screen for character input if you have a PC available).

Today's working environment demands ever-quicker exchange of information, but this is not always easy to achieve. Just think of how much time you spend chasing your colleagues every day.

More often than not, the information you need involves a simple yes or no. SMS is a very efficient way of getting information out to employees quickly, especially for time-critical, brief information. In the business process, brief information is often distributed via e-mail messages that often are read too late. The alternative to e-mail when a quick response is needed is the telephone. But calling people can sometimes be an exercise in futility—they simply don't answer the phone and you find yourself leaving message after message. The best solution is SMS; it can be sent and read from any device, even if the receiver only has access to a cell phone. The main difference between e-mail and SMS is that SMS alerts the receiver that a message has been received, either by way of a buzz or a signal. This way of attracting the receiver's attention is seldom used for e-mail due to the large number of e-mails that are received every day. Therefore, SMS is the preferred function when a quick response is necessary.

When information is not time critical, on the other hand, being able to access it from a mobile device (in the form of e-mail) will make employees more efficient in several ways. For instance, the 10-minute wait outside a client's office can be used to check and reply to e-mail, thereby reducing your office workload. Similarly, you'll be able to send a brief e-mail from the client's site to a colleague if you need additional information—such as the availability of a certain product—and you can get the answer immediately. The key to optimizing the efficiency of mobile employees is to equip them with appropriate mobile devices so that they can make the most of the short time slots that pop up throughout the day.

Filtering

Spam—or unwanted e-mails—is a major cause of irritation to employees. However, people will learn how to deal with this in a timely way. There are two ways of dealing with the internal spam some companies suffer from: education and filtering. Because the use of e-mail has exploded, we cannot expect everyone to understand the negative effects on staff efficiency that unnecessary messages can have. So, organizations will have to train the entire staff in how to use messaging applications. The training will focus on efficiency—emphasizing when and how they should use messaging—rather than on pure technology (which is the main topic in today's courses).

Filtering is an optimal function for people who receive too many messages every day, and filter functions will develop into very smart helpers. Currently, filtering works in different ways, depending on which software you use.

To begin with, you need to understand what a filter is. A filter handles messages based on a set of predetermined rules that identify how messages should be dealt with and defines what action should be taken. If you get many messages where you are listed in the Carbon Copy field (cc), and you get them because the sender thinks you should be informed (you might disagree with this), you probably don't want to have them in the same folder as messages that are addressed directly to you. The filter will simply send the messages in question to a separate folder, and you can read them when you have time—perhaps once a week.

Similarly, you can create a rule that marks all e-mails from one person as "normally not important"; these messages will then be sent to the appropriate folder. Also, if you get external spam, you can mark senders as "junk sender"—all messages from that source will then be deleted directly, and you won't even see them. The reason for sending some messages to folders for later use is that can you get easier access to the important information that sometimes is hard to find in a myriad of messages. As we have seen, mobile devices are excellent tools when you want to optimize the time spent waiting outside a client's office. Dealing with spam is an excellent activity in such circumstances.

Office Applications and Internet/Intranet Access

As we have seen, mobile devices are useful tools in a number of situations. Nevertheless, they do have limitations—you won't be using your mobile device to write a quote while in a cab on your way to a meeting. You will, however, have access to all office tools, which will enable you to view and make minor corrections to material that has been sent to you. You will also be able to create short documents with the mobile device. The functionalities of the Office tools in mobile devices will be more limited than those used in a PC environment—they will, for example, support both the reduced screen size and the smaller keyboard.

When working with back-office applications (your ERP system or client/product information) made available via the intranet or an internet, you will be using communication capabilities that are included in these applications. The information you send will be formatted and entered into your back-office systems and will subsequently be used with other functions at the office. The major advantage here is the ability to access the back-office applications from a remote place.

In most of the examples we have presented, the most tangible advantage is the capability to perform tasks directly. You will no longer have to wait until you get back to the office to update information, edit a quote, or deal with spam. All your information currently is available only on your desktop, but will be accessible through your mobile device—regardless of whether it is in an internal system or on the Internet. The only difference will be the format in which it's presented. You will, of course, also have the choice of working online or offline with the mobile device. Working offline with a mobile device won't entail your having to connect to your PC via a cradle in order to synchronize information, because you will be able to synchronize wirelessly when you want.

Summary

Let's summarize by taking a look at what devices people will be using in the future. You have seen that Office applications are demanding, ergonomically speaking. We have described a number of different devices, ranging from the desktop PC to the PDAs and smartphones. Obviously, you need to be clever when choosing which device you should bring along to different work situations. Even though future PDAs will probably incorporate Office applications, most mobile employees will bring laptops PCs when they need to work with Office applications. It is true that surfing the Web does not require a big screen, but some tasks are simply easier to perform on a big screen than on a small one. Which device people will choose to use is difficult to predict.

CHAPTER

9

Information

Michael Welin-Berger

Consumer Applications

Mobile devices will combine the functionalities of the radio and the Web to leverage customized solutions for users. The synthesis of the information from these two sources will be available on mobile devices. The Web contains just about any information you need—the difficulty is making it easy to find. The more we use mobile devices, the more we will learn about where information can be located. To do so, we will simply adapt the skills we have acquired from the Web. Facilitating this process is the fact that many aspects will be identical to those of PCs—for instance, Web site addresses for content providers will remain the same. The main difference is that we will have several options for how to receive information. Surfing with the mobile device is clearly one way, calling in to a voice application is another, and downloading images and video clips is a third.

Radio on demand will likely be a frequently used service. It is currently available on the Web but it isn't widely used because users are restricted to a PC. In the future, when you listen to your favorite news station on the radio, you won't have to listen to information you aren't interested in; you will be able to select the topics of your choice. The radio will be connected via the Web so that you can choose to listen to the 5 o'clock news at, say, 7 o'clock. The same goes for weather and traffic information. You will be able to download the latest CNN news to your device and listen to it on your way home from work. Needless to say, your device will be connected to cordless hands-free earphones so that you don't disturb your fellow passengers (assuming that you are taking public transport).

Sports, News, Weather, Maps, and Traffic

Sports, news, weather, maps, and traffic are good examples of areas in which mobility can add real value. You will no longer be restricted to the place where you usually connect (for example, your home PC). You will also be able to choose when and where you receive the latest news. We will now briefly describe how mobile functionalities can be used to access content that traditionally has been accessible only via fixed devices. The basic premise is that there will be no limit to your options because any kind of information you desire will be available to you—at any time or place.

Sports

Sports attract the interest of a large number of people. Their passion drives them to delve into the sports pages of the paper first thing in the morning or watch the sports channel as often as possible. Many people also spend a great deal of time in front of their computers to get the latest news on their favorite sports. The mobile device will be a great new channel for people to get detailed and rich information at any time. Local sports events that are broadcast on the radio in any given city will be made available to the whole world thanks to the Web. And if the news is available on the Web, it will also be accessible from mobile devices. The real-time experience—seeing things as they happen—will bring tremendous value to communication.

The cost of covering an event will be significantly reduced as a result of broadcasting on the Internet, irrespective of whether you choose video streaming or mere voice broadcast. You'll be able to follow events that are currently not covered by the media due to an insufficient number of spectators (spectators play an important role in generating the money needed to cover broadcasting costs). The Internet will make it possible to broadcast such local events as tennis games or golf tournaments via voice. Only one person will have to cover the event, making the overall broadcasting cost very low. With larger commercial events such as the professional golf tour, broadcast video will be used so that everyone can see the event—even from their mobile devices.

News

Traditional news services will evolve into customized news services adapted to users' specific needs. The content will be much richer because video will be used in combination with text and voice. This means that you will no longer be dependent on TV networks for visual news. You will also be able to indicate which geographical areas you are interested in and the news will be provided visually—for example, by using pictures and video sequences.

Let's look at some examples of this. If you're interested in the European economy and American football, the first news page in your device will focus on those topics. You will receive information from other areas only if something out of the ordinary has taken place. While viewing your "info page," you will be provided with links to further information related to what is described in the headlines. It is very much like CNN on the Internet today, with the important distinction that the first page only displays the content you are interested in. Furthermore, you will have access to much more customized information in your chosen areas—information you otherwise would only find in the trade press (for example, business, sports, medicine, politics). Some of these services will demand some kind of paid subscription, while advertising will finance others.

Weather

Mobile devices will dramatically change the presentation of weather forecasts. You will automatically get forecasts covering the geographical area you are in (as default). This will be enabled by localization/positioning, whereby the weather service can identify your exact location and provide you with information regarding your surroundings. This is a major improvement on traditional TV forecasts with their country/city maps. If you like the outdoors, you will have the possibility of receiving customized weather reports that provide you with information on the latest sailing or skiing conditions. For operators, the key factor is to adapt the available information to the needs of the viewers/listeners. Localization/positioning will be provided by operators, and all you will have to do to get the latest information is to make a selection—for example, sea, snow, or ordinary weather conditions. Operators will know which base station you are connected to and will therefore be able to provide you with both the service and the information.

In Chapter 10, we will describe how these services will be used in the enterprise environment.

Maps

Wouldn't it be great if we could have a detailed map of the world in our pocket? Well, that will definitely be possible. Today, maps are either available in paper format or via a PC on the Internet. As a result, we rarely have access to them when we need them. We will now have access to all Yellow Pages services in our PDAs—all locations will have route descriptions and maps on their Web sites. A driving instruction functionality (which already exists in many cars) will also be available on mobile devices. So, you are when walking down the street trying to find a restaurant, you will have access to the same map functionality that many people have in their cars.

In the high-end segment, cars have had built-in driving directions for some time. Taxis have access to driving information in the form of a CD containing all relevant map information. This is a relatively expensive solution, and you have to upgrade your CD if you want to keep abreast of all the changes made to the available road systems. In the future, we will have applications connected to our mobile devices that give us this information (including maps) wirelessly. This will also take care of the problem of built-in devices in vehicles, which have a different life cycle than mobile devices. A car with a built-in device can run for about 10 years, whereas a traditional mobile device becomes outdated after three years.

In conclusion, map functionality is a good example of the different kinds of services that mobile devices, such as PDAs, excel in providing. You will not buy a device solely for this functionality, but when you have the device you will probably use it extensively.

Traffic

Drivers will be updated on traffic situations in a much more detailed way than today. The content will depend on where they are and what kind of information they want. The reasoning here is the same as that behind news and weather content: We will be able to define what kind of information we want and the system will be able to identify our geographical location at any point in time.

Today, traffic announcements on the radio are sent to every listener in the city. With the new mobile functionality, it will be possible to access information that is specific to the route you are on, in addition to important events developing in its proximity. You won't, in other words, have to listen to traffic-jam coverage on the north side of the city when you are driving on the south side. You will also be able to listen to local traffic updates when leaving your driveway, which will help you decide which route you are going to take to the office.

Mobile devices in cars have, until now, been restricted to such support functions as driving directions and security functionalities (for example, an alarm that pinpoints your position in emergencies). In the future, however, you will have access to additional functionalities that enable a variety of services based on the geographical area of your choice, including news, weather, and sports. What's more, the information will be provided when you ask for it—so, the music you are listening to won't be interrupted by traffic updates that aren't of interest to you.

Entertainment: E-books, Music, Gambling, and Multiplayer Games

There are very high expectations within the entertainment arena for mobile services. There is an opportunity for tremendous growth in mobile services, assuming that service providers manage to offer the services users really want. One difficulty is to balance the launch of new services with the development of new devices; users will demand services that are user-friendly and not as complicated as WAP (Wireless Access Protocol) was initially.

In terms of entertainment services, WAP (which is text-based) has limited functionality. The only services of interest it can provide are sports results and information about where movies are playing. Users will want mobile devices that offer more services and, perhaps more importantly, they will want them to have a graphical user interface. This, in combination with the improved availability of these services, will help to create a large market for mobile services. With the increased bandwidth available in approximately 2004, it will be possible to listen to music, see movie trailers, and download books to mobile devices. In fact, they will also allow you to watch TV. The screen probably won't be good enough for viewing wildlife programs, but you will be able to watch the news, talk shows, and quiz shows.

E-Books

The Internet will be your library! You will be able to access any book you have read or want to read without having to go to the bookstore. If you feel like reading a novel written by a specific writer, all you will have to do is connect, pay, and download the book directly into your device. This can be done while you are on vacation anywhere in the world. The purchase of books has become simplified thanks to the Internet, but making books directly available to you is a huge step for the book industry. The availability of e-books will create fierce competition among Web-based and traditional bookstores. We will, of course, occasionally still want to turn the pages in a real book. But the convenience of direct access when we want something to read will create a demand for e-books.

There are different opinions about the future of books. Some people believe that in the next 5–10 years, most books will be sold as e-books that are distributed and used electronically. The term *e-book* refers to all mobile devices aimed at providing the user with information in the same way that we use books today. It will probably take some time, but traditional books will, in the future, be considered a luxury item, much as we today consider hardbound books to be nicer than paperbacks. For students, e-books will be a blessing, because they will cut education costs. This also applies to all reference literature we will need from time to time—the e-book is the perfect tool in this respect. It will, however, take us longer to accept the e-book as entertainment on a trip or as reading on the beach due to people's deeply rooted habits. There are other obstacles that the e-book must overcome. We will demand better screens that don't create eye strain, technology that will take some time to develop. Such screens already exist, but they are very expensive and consume a lot of power, which reduces battery time significantly.

Music

You will be able to choose between thousands of radio stations when you connect your mobile device to the Internet. All radio stations will broadcast on the Internet, making it possible for anyone in the world to listen to them. You will, for example, be able to listen to Albanian folk music, if that's what it takes to cheer you up. This functionality will change the entire business structure of radio stations.

The possibility of downloading music from the Internet has been around for a while now, and the usage is increasing. Lately, we have seen new formats for storing and distributing music emerge, and their market share is increasing rapidly. MP3 is the most common format currently in use besides traditional CDs. MP3 is, however, not only useful for storing music on a portable music player—it can also be used to download music directly from the Web to a device. The key feature is to compress the size of music files without losing too much quality, and MP3s are a good compromise in this regard. Of course, we will see new and improved formats in time, but MP3 will be the format of choice for some years to come. This means that you can connect to your music library on the Web and choose the music you want to listen to at any particular point in time, regardless of whether it is available on a CD at home or not. A compressed format will be as essential as bandwidth to the delivery of music—in terms of both increasing quality and reducing download time.

Gambling

Gambling will be one of the so-called "killer applications" of mobile devices. The reason for this is that the potential market will drive existing gambling institutions to invest in a number of different gambling options. New kinds of games will appear—you will be able to play poker against other people by using your mobile device. However, the largest market is expected to develop in traditional gambling activities such as horse racing, football, and soccer. Here, the main task will be to capitalize on specific situations in which players don't have access to a PC. To do so, gambling companies must extend their existing Web gambling services to mobile devices.

To illustrate how mobile devices can be used for gambling purposes, let's take a look at horse racing. Mobile horse racing gambling provides you with two new features: placing a bet at the last minute and doing so from any geographical location. The first feature supports the gambler's desire to get as much last-minute information on the horses as possible. This information, with tips from specialists, is currently available on the Internet, but it's not available to players when they are at the track or away from a connected PC. Getting access to statistics on the horses' latest results or information on how the warm-up went could prove very valuable.

The second feature, placing a bet from virtually anywhere, will also enhance the gambling experience. The possibility of placing a bet even though you are on a business trip and can't make it to the race track will increase gambling activity substantially. There will be no need to go to the ticket window to place your bet—even when you're physically at the race track—because you will be able to do so from your mobile device. Existing payment structures will simply be extended to the mobile device, thereby making the cost per transaction very low.

We chose horse racing as an example, but we could just as well have picked basketball, Formula 1, football, soccer, or anything else—the same functionality is applicable everywhere. Needless to say, mobility will bring a new dimension to the gambling experience.

Multiplayer Games

Multiplayer games are expected to tap into a huge market among young people worldwide. The reason for this assumption is the experiences we have from mobile devices in Japan, SMS communication via mobile phones in Europe, and gambling via PCs on the Internet. Young people take mobile devices to heart and are prepared to spend substantial amounts of money for this functionality. In Japan, for example, young people play both traditional games like Super Mario and advanced role-playing games with several participants.

We have seen the Internet create a new kind of activity in which people—mostly youths—play different games against each other on the Web. The most common ones include strategy games, chess, and dominos. Many of these games can be played on a relatively small screen and will be played from different terminals, such as connected PCs and mobile devices. It is difficult to predict if these activities will grow in mobile use or if they will just fade out due to the availability of bigger screens at home.

The future development of games will also include new functionalities such as positioning. We will see games where people compete at finding each other in a city without being discovered themselves. Again, it's difficult to foresee what these new games will look like, but we will certainly see completely new games evolve as a result of new functionalities in mobile devices.

Education

Mobile devices will significantly improve our knowledge in several areas as well as change the way we learn to use different kinds of equipment. They will enhance access to educational material such as manuals and textbooks. The key feature will be to have the information available at all times from the mobile device, which can be consulted when discussing something with your friends or at work. Alternately, if you are having difficulties with your boat, car, or motorcycle, you will have direct access to a comprehensive manual.

The mobile device is not an optimal tool for general, time-consuming searches on the Internet; you will probably prefer to do this from a PC with a big screen. However, when you are in need of specific information, such things as manuals and textbooks will be easy to access and use. Let's look at some examples. Imagine that you are a lawyer discussing a legal dilemma with a colleague and you realize that going through the law book to find the answer will take some time. You do, however, recall that pertinent information on the topic could be found in a reference book that you don't have readily available. Because you know which book you are looking for, finding the chapter and refreshing your memory will be relatively easy.

This also applies to the manual delivered with your car, which is very concise since it has to be small. Let's say that you have blown a fuse in your car. The manual in the glove compartment is not detailed enough to help you solve the problem. You need more detailed information and instructions, so you simply access the comprehensive manual in electronic format (which has been provided by the car manufacturer) via your mobile device. This manual, unlike its sibling in the glove compartment, is not limited in size and is therefore a much more helpful reference source.

Emergency and Disaster Services

Ambulances, fire brigades, and different kinds of command vehicles used by the military currently make use of many mobile applications. Ambulances are being equipped with devices that both enhance the level of care en route to the hospital and prepare the hospital for what is needed as soon as the ambulance arrives.

Cardiology is an area that has been prioritized due to the importance of rapid treatment when someone is suffering from a heart attack. The time that elapses between a heart attack and treatment is critical. Another reason why mobile devices can be helpful in cardiology is that it demands a great deal of specialized knowledge and extensive experience. Mobile devices will support communication between the doctor at the hospital and the emergency medical technician in the ambulance, as well as supporting functionalities in the ambulance itself. Instructions and checklists will instantly be made available and we will see much more sophisticated equipment.

Mobile services can also be applied to *fleet management* in connection with emergencies and disasters. Positioning units and assessing their availability will increase efficiency and probably save lives in the future. Fire brigades mostly rely on voice communication between available units when fighting forest fires. This is changing rapidly to enable them to manage their operations more effectively; one of the functionalities will provide staff with maps and instructions through mobile devices.

Enterprise Processes

Besides the access to the back-office applications, there are many consumer services that will be used by enterprises as well. We will look at news, weather maps, and traffic as examples of this.

News

Businesspeople place high demands on news services. The feeling of being on top of things and knowing what is happening in the world drives them to read business papers in addition to newspapers. Many business people also subscribe to different news services that cover their particular business sector or a specific topic of interest. These services are distributed via the Web, mainly as e-mails. The problem is that the information ends up together with all the other information you need to deal with in the office. So, despite the fact that you subscribed to the service with the best of intentions, you rarely have the time to read the articles. If you could access these e-mails from a mobile device, however, you could make use of any time slots that might pop up in order to update yourself.

Many countries have already been testing this solution in the financial services area. The obvious need for up-to-date information has driven several initiatives that until now have been held back by a lack of technology. Most services have been SMS/WAP enabled or synchronized solutions with "old" information that is available to you after disconnecting. We will see rapid expansion in this field, with new devices that are easier to read and significantly easier to connect to different services. You will be able to read abstracts and download entire articles on anything that interests you. This will be applied to all sector-oriented services that reside on the Internet.

Mobile devices that provide you with access to the information you want, when you have the time to read it, will change the way you gather information. This also applies to internal information, which can be distributed via an intranet. Most employees want to update themselves but rarely find the time to do so. Now they can do it when it suits them.

Weather, Maps, and Traffic

The transportation industry is the most frequent user of today's mobile devices. There are two main ways of distributing information, but neither one is very good. The first is through the radio, which broadcasts information continually regardless of your specific needs. The second is through the Web, which has all the necessary services available but isn't accessible to the remotely located person (for example, a truck driver). Taxi companies, for example, send information to their fleets via mobile devices as soon as important events take place. But this information needs to be disseminated with care—only cars that are directly affected by the information should receive it (that is, cars servicing the same region).

The perfect service would be to give drivers access to all the information that is currently available on the Web—weather, maps, and traffic. These services, which are available in mobile devices, will reduce the need for support from the dispatching center.

The reason there is anticipated growth in this area is the similar needs employees experience in many different companies. Content providers have a hard time financing their businesses today, because there simply aren't enough enterprise customers that use the existing functionalities. With connected mobile devices, however, the market will grow. The need is there and this will create great business opportunities for different content providers.

Content providers are already collecting the information and that is costing them money. This cost will increase when companies demand customized solutions. Customization will be provided based on such parameters as geographical location or areas of interest. The main difference is that there will actually be a customer who is prepared to pay for these services, which they believe will contribute to their business.

Management Dashboard

Let's start this section with a definition of management dashboard:

> *The management dashboard refers to a functionality that will allow managers to have access to all relevant information that resides in internal systems, as well as all relevant external information (such as financial or sector-specific information).*

More often than not, decisions are based upon incorrect assumptions because, at the time of the decision, certain information was not available. When you meet with a customer, many decisions would be different if you had access to all the information that today is only available to you on your desktop computer. Information always plays a key role in making good decisions. In the future, information will be available to you by giving your mobile devices the functionality to receive the necessary information. That way, you will have access to it whenever you need it. Some suppliers call management dashboards "decision support systems" to help explain this rather vague concept.

The need for information on a frequent basis and occasional information needs will both be met by different technical solutions. Information that you need frequent access to will be synchronized with your mobile device and stored for immediate access, whereas all other information will be accessible from the device just as it is from your desktop computer. Thus, you will have access to all reports, sales information, stock updates, and the like within a matter of seconds. Imagine that you are in a management meeting and you are discussing different cost-cutting strategies. The capability to get connected and instantly receive all relevant information (for example, the latest financial results) will definitely increase the quality of decisions.

We can already see that all big ERP and CRM (Consumer Relationship Management) application suppliers have started to create mobile extensions to their systems. These extensions provide access to the application from different devices. However, they are currently focusing on functionalities for mobile technicians or sales reps rather than on supplying management with mobile functions. Application suppliers also have a problem with their technical heritage, which demands considerable changes in core IT processes linked to the distribution of information.

Thus, we see the same slow evolution as when ERP and CRM suppliers began to adopt Web technology as a way to communicate. It will probably take several years before they have reached the same usage rates of mobile devices. As a result, many companies will start building "homemade" applications for their own specific needs. They will combine back-office information with new information that has not been relevant to today's stationary use (for example, driving instructions, weather, and news).

One of the great advantages with the technical evolution caused by the Internet is that the interfaces with the back-office systems are now standardized. Mobile functionality will therefore be relatively easy to integrate with other solutions residing in the back-office because they will use the same interfaces.

Summary

The key issue when designing the usage models in an enterprise involves reusing existing functionality from the consumer space. This gives the service provider an incentive to adjust services to the enterprise space. The second step involves determining whether the service can be adjusted to your needs; this is becoming more common with service providers who are aiming for the enterprise space as well.

When you have established the usage model, you then need to determine the specific services available for your sector. This issue is covered in Chapter 10.

CHAPTER

10

Commerce

Michael Welin-Berger

Applications in Different Sectors

When addressing how you will use new mobile tools, we have to combine the old services with the new possibilities. The goal is to help employees and customers to be more efficient by giving them more choices for how they want to use these services. It is normally a good starting point to make some existing Web services available on the new mobile devices.

It is apparent that companies that used the Web early on were also among the first to embrace mobile communication. They learned the lesson that it takes more time than expected. To begin with, there was an overwhelming focus on the external marketplace, but most of the initial tangible benefits actually came when companies made all of their internal information available on the intranet. We will see a similar development in the mobile arena.

Communication with Your Clients

If you use the Web when communicating with your customers today, you have probably learned that having a solid functionality is more important than using the latest technology. Therefore, the important question is: How can your users benefit from having a service delivered to a mobile device? This is indeed a more important query than the more common one: What can be delivered? There are too many technology-driven projects today that focus on devices rather than on how to improve user's functionality. For this reason, it is important to clearly describe the tangible benefits to clients. The sooner the better, because it takes time to change people's habits.

Here are some examples of existing services:

- **Airlines that provide travel updates via mobile devices.** This is a perfect service for customers on the move who need to know whether a plane is going to depart on time. The same concept can be applied to train travel.
- **Booking movie tickets.** Going to a movie is very often a spontaneous decision, so this is a good example of how mobility can generate added value. Customers cannot be expected to know which movie they want to see—you have to sell the movie to them. This is being done on the Web today. For mobile services, the biggest challenge is to show short trailers that will help customers in their choice of movie.
- **Providing the delivery status of goods being delivered to a customer combined with expected arrival at client site.** There is usually a great deal of information within a company's internal systems. It simply needs to be made available to the customers (in limited form). We have seen how UPS and DHL deliver information on the Web; this feature is now being extended into mobile devices.

The important place for any organization to start is with its customers through the use of mobile devices—probably with applications based on existing Web services—and then add new services. For companies that do not have any existing customer-targeted Web services that could be enhanced by mobile solutions, the best plan of action is to start internally.

Advertising

Advertising on the Web has not been a great success financially speaking and many might argue that advertising via mobile devices will not be profitable either. However, one important dimension will be introduced thanks to mobility: *positioning*. We now have an additional piece of information about users; their position will in many instances help us pitch the potential client group for an ad. Let's look at an example. A restaurant sends an ad to users who are within a 500-yard radius, in which they offer a three-course meal for less than $15. This would probably attract some new customers. We do not know today if users will allow positioning to be used for commercial purposes. If they do, providing nearby customers with reduced rates or other special offers could potentially create a substantial user group. In short, apart from positioning, there is no significant difference between the mobile user and the traditional Web user when it comes to advertising. The only thing worth mentioning is the smaller screen—due to the size of the screen, ads cannot be very large. But size is not important if you can create the right kind of information.

The key to creating truly valuable information—instead of annoying disturbances—lies in finding user groups that have a specific area of interest. The easiest sector to apply this to is one that already exists—the music industry. Here, customers are keen on keeping abreast with the latest releases. For example, if you take a community of Madonna fans, there is great potential for sending them updates as soon as she releases a new album. This would also be a good marketing tool for new albums; short excerpts from new songs could be disseminated in a marketing effort prior to each album's release.

This kind of activity could also be combined with access to images and the use of the artist's voice on your answering service. Your friends might hear Madonna's voice in your answering service saying "Michael can't take your call right now, please leave a message."

Another community that might benefit from value-added information is readers with a specific area of interest, such as a certain author or a particular genre. Being informed that there is a new novel in their area of interest available—with a link to a summary—will doubtless be regarded by many users as a value-adding service.

Yet another area that lends itself to mobile services is sports. There is a huge interest in getting the latest results or mid-game scores, and there is an endless amount of information that you can tap into. Ads that appear in connection with sporting events do, however, need to be presented discretely. Customers should not consider them a nuisance. One solution is the form of advertising often seen on scoreboards (for example, "scores provided by IBM"). The most efficient kind of communication will be short and combined with images—the use of logos will further enhance the value of a good brand.

In all these cases presented, mobile services will make it possible to send instant information to existing communities that have a clear interest in obtaining such information. It is essential that these mobile services be integrated into a company's customer-communication strategy, much in the same way the Internet is now used. There are a plethora of communities that can be targeted, all of which have distinct needs. Some groups, such as youths, will want to know everything about Eminem's latest album. The sooner you know about it, the cooler you are. Others, such as avid sports fans, might want to know the latest news in the world of Formula 1.

Financial Services

The increasingly competitive nature of the banking sector has played a major role in the recent development of direct distribution channels, such as the Internet. For existing banks, it's imperative that direct customer relationships be maintained and enhanced, and that the products and value-added services be delivered at a minimum cost. That is the basic reasoning behind the expected fast adoption of new channels for communication with clients.

The financial services area began with different services for smartphones. In Europe, most applications were WAP-based, whereas Asia had several applications that were SIM-based. These early projects have not been very successful—probably due to a relatively complicated user experience. Asia has been more successful than Europe in this respect, most likely as a result of their applications being more user-friendly. New security solutions are now being presented that can provide users with a different experience: they can have a graphical user interface in the device. For this reason, we will separate the discussion into two technology streams: extended functionality on the phone (using the SIM card) and browser-based application in PDAs. These two streams will subsequently be used for different purposes in different markets.

Description of the Customer Value Proposition

In customer value propositioning, the basic concept is to offer customers a range of banking and related value-added services via a cell phone or PDA. The objective is to provide easy access to banking information for customers on the move, as part of a portfolio of electronic channels.

The services that most likely will be used early on involve simple extensions of current Internet banking services, such as balance and credit card inquiries, money transfers, foreign exchange information, bill payment services, the purchasing of mutual funds, and stock-market information.

These services are predominantly "pull" services. *Pull services* are services where customers actively demand (or pull through) and take the initiative in executing an action. "Pull" services will be combined with different "push" services, in which banks actively deliver information to their customers based on prearranged, customer-specific profiles. Examples of push services are stock alerts, portfolio information, and other informational services.

Besides functionality, there is the issue of who should be targeted for push and pull services. We need to define which groups would most likely use these services. Here, we can return to the same criteria used to identify early adopters of current Web services: people between the ages of 25 and 45, high-income groups, and early adopters of new technology.

Expected Technology Evolution

We expect rapid growth in the use of mobile devices. However, other kinds of technology need to be applied in combination with those currently used in most applications (such as video). The most widely used technology will most likely be a browser-based functionality on a PDA. Therefore, we need to view the functionality from two separate technological perspectives: phone-based applications using SIM cards and browser-based applications for PDAs. These two choices do, however, introduce separate issues. The first is how to deal with security, which will be handled differently in the two devices. The second issue will be to adapt the presentation of information to the user interface of the devices.

Phone-Based Applications Using SIM Cards

A digital certificate located in the SIM card will enable a high level of security. This also makes it possible to provide some of the software on the SIM card, which in turn increases the speed of the application. One downside to SIM-card solutions is the poorer interface provided by the phone—currently in the form of WAP or similar solutions. In the future, improved interfaces will be available (they will be delivered with Java applications, which will enhance the user experience). However, the main drawback is that these solutions require that information resides on the SIM card—and that information can only be administered by the card issuer (that is, the operator). We will perhaps see devices with two SIM slots—one for the operator SIM card and another for the security-providing SIM card. This demands a coordinated approach from device manufacturers and application providers, and it is hard to predict whether dual SIM-slot phones will ever be launched.

Browser-Based Applications for PDAs

Browser-based applications for PDAs will make use of security functionalities that already exist on the Internet. This will make it easy to implement the applications as soon as devices are regularly used. Browser-based applications will not require any extra administration or implementation of additional software. Consequently, the rollout of these mobile services will be very fast and that will quickly give rise to a substantial market. The interface will be similar to that of a traditional Web browser—it will simply be adapted to fit smaller screens.

Manufacturing

Manufacturing is a large sector containing a wide variety of enterprises—from manufacturers of heavy industrial equipment to makers of small electronic devices for the consumer market. Manufacturing is in need of a generic functionality for sales force and field service automation. (*Field service* refers to all business processes that support the use of a product or service on-site.) Here, we will briefly describe how these functionalities can be used.

Sales force automation needs are dependent on how sales are made. Although sales are usually made through retailers, sales force automation can also be applied to enterprises that sell industrial equipment or the like directly to an end customer. The need for automation will be the same, regardless of what is being sold. Sales reps will have direct access to all pertinent product information that today is available to them only when they were sitting at their desk (for example, client information, recent contracts, the supply of goods, and notes from the last meeting). Here, added value is manifested in better service, improved sales rep efficiency, and reduced back-office administration.

Field service automation is related to the service and repair of equipment outside the enterprise plant. The need is similar to that of sales force automation and can be summarized as having direct access to all the information available when you are sitting at a desk (for example, client information, information related to equipment delivered to the client, and earlier work performed on that equipment). This applies to all equipment that is not delivered to a service center when service or repair needs to be performed. The added value generated by field service automation will be seen in improved efficiency for the technician and reduced back-office workload.

There are two areas specific to manufacturing where mobility brings tangible benefits. The first area is the connection of equipment to remote control centers. Mobile devices will most likely support forestry and mining equipment as well as specially designed trucks. The reason for this is the complexity of the equipment—there are several functions that can be enhanced with the help of mobile devices. Let's look at a forklift to further illustrate this. In many plants, forklifts can be monitored from a distance to obtain information on mileage and oil levels. This means that the forklift has to be equipped in such a way that it can provide the desired information. It has to have built-in support systems that can be connected to a remote control center. The forklift also has to be equipped with a rugged, well-positioned screen that will not affect the drivers when they are working. If the equipment doesn't come with a screen, there will be the possibility of connecting it to a mobile device, such as a PDA. The PDA will then act as the interface for adjusting settings or locating malfunctions.

The second area is the bundling of mobile functionalities with sold equipment to improve performance, such as combining heavy equipment with a PDA for wireless monitoring. New mobile functionalities are constantly evolving, and there is a normal evolution of this functionality within an organization. First, the technician becomes connected to the back-office to report work orders and access all relevant information about the equipment in question. Second, the technician can connect the mobile device to the equipment, a PDA, for instance. This enables the technician to adapt settings or monitor performance while standing next to the equipment. Third, the equipment becomes connected remotely to some kind of surveillance center that monitors and adjusts performance from a distance.

The car is a good example of the complex functionalities we want to have built into a device for our convenience. The experiences that car manufacturers capitalized on when designing their cars, however, came from truck manufacturers, because they were the ones who started implementing wirelessly connected devices in vehicles.

When we look at the new requirements on a vehicle's infrastructure, we can see different needs that have to be addressed in different ways. New high-end cars have three separate networks built in from the outset. First, there's the network for the engine, the brakes, and all other aspects that are essential to the performance and security of the car. Second, a network controls such things as windows, lights, radio, and the AC. Third, you have what could be called the multimedia network, which enables you to communicate through music, e-mail, voice, positioning systems, and a gaming environment for the kids in the back seat.

Telecom Sector

In a discussion about mobile functionalities, you can't avoid addressing telecom operators. Mobility is, naturally, important to telecom operators because they are the providers of both voice and data access. But there are other areas outside the provisioning of a mobile infrastructure that are of interest to telecom operators.

One of the areas with highest expectations was—and still is—*micropayment* (a purchase under one dollar). The reasoning behind this is twofold: there is a well-established payment relationship between telecom operators and their customers, and sophisticated systems are already in place to handle billing in different ways. It was expected that telecom operators would enter the market much as NTT DoCoMo did in Japan. They became the channel for mobile payment (of course, they charged a fee for these services). NTT DoCoMo started out by charging 9 percent of the client revenue; in Europe this has been judged as being too low a fee. This is one of the most astute decisions made by NTT DoCoMo. They chose to set a low fee for administrating the payments, thereby enabling a large market volume rather than fighting for a higher percentage of a smaller volume.

Another interesting area is *positioning*—a technology that provides operators with information about where their subscribers are as long as their cell phones are switched on. Operators can locate your position to within a 200-yard radius. Positioning does not demand substantial investments for operators, and customers want this service. There is, however, a need for a speedy rollout because GPS (Global Positioning System), a competing technology, will soon be built into mobile devices. GPS positioning is far more accurate than base-station positioning—it can pinpoint a subscriber's position within 1 yard. If operators could provide a base-station positioning service now, it is likely that many companies would want to incorporate it in their employees' mobile devices. If operators wait too long, however, the preferred choice will be GPS. There is actually a third technology that makes it possible for operators to provide a positioning service that is almost as accurate as GPS positioning, namely *triangulation*. But triangulation is a relatively complicated technology that demands contact with two or more base stations simultaneously.

Internally, telecom operators use mobile devices mainly for field service and sales force automation, albeit with a focus on field service automation. There are a great number of radio towers and other systems related to the mobile infrastructure that have to be maintained and repaired on a regular basis. This work must be carried out by technicians who need access to information about equipment while working on-site. These are very much the same needs as the ones we have described in other sectors—access to back-office systems for equipment information and reporting on work that has been completed.

Media Sector

The media is a sector that has tried to capitalize on the rise of the Internet. A number of newspapers aggressively marketed their Web-based editions. They tried to tie their readers closer to the paper by providing content on the Web and later on mobile devices. The Web version didn't always include the entire content of the paper version, and mobile devices were understandably used as a condensed news service. AvantGo (www.avantgo.com) is one of the more successful providers of synchronized content from different newspapers to mobile devices.

The reason for the early success of synchronized solutions was the lack of good wireless access. It is easy to use—you simply connect a PDA to your PC via a cradle. When you bring a PDA with you, all the information is downloaded into the device, and you can read the content anywhere you want to without having to be connected. With the new wireless networks, synchronization with the PC will be replaced by a functionality that allows you to synchronize the information even when you are away from your PC. This will be made possible through the use of wireless networks. We will, of course, continue to use browsers to access online information as well, but the basic information need from different sources will be synchronized with the device.

We will see a gradual shift into content-rich media services that combine news from papers and video clips, much like CNN's Web service. Advertising is often the main source of revenue for media companies. We have already described some of the advertisement possibilities that will result from positioning services. The key issue for media companies will be how to combine traditional ads with advertising on mobile devices.

The greatest opportunity for news services lies in targeting different communities with information adapted to their needs—that is, designing the information differently for different user groups. The combination of newspapers, the Web, and mobile channels will meet different needs within a user group. This has to be combined with customized information, and that means sending the desired information to the customer's device of choice. Furthermore, when using a particular device, you will already have specified exactly what information you want every time you connect to a service. For instance, if you're more interested in finance than in politics, the service will present you with customized information, and you will receive valuable information about finance and only condensed information on the most important political events. This is a challenge for those media providers who want to benefit from mobile channels.

Retail Sector

In the retail sector, there are two main ways of providing mobile services to the customer, namely product registration and payment.

Let's start with registration. We often find ourselves standing in a line at the supermarket, vowing never again to go grocery shopping on a Friday afternoon. A new service is currently being tested around the world: direct registration of any product you put in your shopping cart. By using a mobile device equipped with a bar-code scanner, you can register items as you pick them off the shelf and then pack them directly into the paper bags in your shopping cart. When you've completed your shopping round, you simply pay the amount that is displayed on the mobile device—there is no need to go through a traditional cashier. This is a good solution for all parties. The only thing that supermarkets have to do on a random basis is to inspect that customers actually register all the items in their shopping cart. When tested in Sweden, a system was adopted whereby customers who have passed earlier inspections are checked less frequently. They are labeled as "trustworthy" in the customer information database.

The other interesting area is payment. Large department stores will be at the forefront of providing their customers with new payment methods. Mobile devices are optimal tools for making smooth and efficient payments. They will be connected to the cashier in some way—either through Bluetooth, infrared, or through a call from a cell phone.

Mobility can even be used to enhance internal logistical processes within retailing. To illustrate this, we will now look at the processes inside warehouses and shops.

Mobile devices have been used for several years in warehouses, mainly in the form of specialized handsets and terminals on forklifts. With new technology, it will become cheaper and easier to connect to other systems than the systems that administrate the logistics flow. The reason the cost will be lower is that we will be able to use standardized equipment to a greater extent, which in turn reduces both application development costs and the cost per device. It will also be easier to integrate other functionalities, such as maintenance information, messaging, and other internal procedures. We will see a similar development regarding the workforce in shops. Currently, if paper is not used, orders are placed by using a specialized device that is connected directly to the purchasing system. The new devices will also provide direct access to the supplier's informational sites, where availability and different campaigns can influence what consumers are buying.

Healthcare and Life Sciences Sectors

In the healthcare and life sciences sectors, internal processes will be made more efficient as a result of mobile functionalities, which will be used to enable care outside the clinic, support the sales force, and even speed up clinical trials.

Healthcare will adopt mobile functionalities, and they will be used extensively both inside and outside of hospitals. Today, the lack of access to patient information when patients are cared for outside the hospital gives rise to many treatment errors, because all pertinent patient information is not available to the doctor. In the future, with the help of mobile devices, all staff providing care outside a hospital—be it at home or at a nursing home—will have access to the same information as that available in the hospital. This includes all the procedures that are related to out-of-hospital care, such as planning treatment and making checklists for different categories of patients.

Security is an important issue in the healthcare sector. There is a strong demand for security and there is concern that the distribution of patient information to mobile devices does not meet security requirements. But that is not the case. When a doctor makes his rounds inside a hospital, he will be able to access patient information from a mobile device—even if he is away from the ward. Furthermore, there will be no need for administrative staff to handle the prescriptions afterwards because everything the doctor prescribes will be entered directly into the system.

From a life sciences standpoint, we will see extensive external communication during clinical trials. Here, mobile devices will definitely play a major role. Clinical trials, which are normally performed globally in many hospitals, are enormously costly to perform. Anything that speeds up the process and/or enhances quality will result in substantial cost savings. Getting away from paper forms will be of interest to all organizations working with clinical trials. But reduced paperwork won't be the only advantage. Mobile devices will enable doctors and nurses to quickly access any information provided by a patient during a trial. Thus, if a patient is experiencing side effects, it will be easy to access his or her patient information and then make a diagnosis. Yet another reason why mobile devices are preferable to paper forms is that doctors and nurses will get instant feedback in the event that misinformation has been entered into the system.

In this sector, the adoption of mobile devices should be rather quick because the end-user has an academic degree and is usually relatively computer literate.

Voice applications that are directly connected to IT systems will improve efficiency when information is desired from a large group of patients. Today, when a drug has been approved and has started being used by patients, staff members collect information by calling patients to ask them questions about the drug. It would be much more efficient if patients could simply call in and register the information themselves. That way, only patients who have forgotten to submit information need to be contacted. Voice application is also easy to use for elderly people who have very little experience in using computers and cell phones.

The life science sector has a unique situation when it comes to sales. The actual purchase of a product is performed after the meeting between the doctor and the sales rep, a meeting that can be seen as more of a brief marketing campaign. The availability of product information is crucial in these meetings—a quick answer during the meeting is much more valuable than having to get back with an answer via e-mail the next day. This can be done with mobile devices equipped with a knowledge-based application designed to answer questions from doctors.

Energy and Utilities Organizations

Energy and utilities organizations have a huge potential to gain benefits from mobile solutions. The main benefits will be in the form of enhanced internal processes and increased communication with clients. Field service is the area in which most efforts are currently being focused, and here we can see considerable gains in the projects that are being implemented. There are a great number of power lines and other systems related to the infrastructure, and they have to be maintained and repaired on a regular basis. This work must be carried out by technicians who need access to information about equipment while working on-site. We will also see increased usage of remote monitoring, because the infrastructure is already being in place to provide communication to the remote systems.

The workforce in these sectors deals with equipment in two types of remote sites: smaller sites that handle the distribution of electricity or water, and larger sites where electricity is generated, such as power plants. When working at smaller sites, a PDA connected to the back-office systems will provide staff with the necessary information. When working on larger sites, however, all equipment will be connected to remote monitoring that will provide information on the status of the system and any necessary adjustments. Remote monitoring will not reduce the need for mobile devices because the equipment will still have to be repaired away from a connected PC.

Meter reading is one area that will change dramatically. Currently, utility companies usually send out subcontracted staff to check meters. The process of writing out the consumption on a paper and then entering the values into the system causes errors and is time-consuming. The first step will be to connect the electricity meter to a mobile device, which then can be connected to a back-office system. After that, meters will be directly connected to the central systems and billing will be calculated using up-to-date information.

Commerce and Enterprise Processes

The rise in the use of computers in companies has led to great efficiency improvements. Even though the paper-free office is not here quite yet, we have come a long way in terms of reduction in the use of paper in core processes. This does not, however, apply to the mobile workforce. Field engineers often receive printed work orders providing information on all the tasks that need to be performed and parts that will be needed. Work orders usually come in four copies: one each for the client, the engineer, the back-office, and filing purposes. Improved speed and reduced back-office workload will be two important arguments for introducing the concept of making work orders available through mobile devices.

Internet technology is used to enhance processes in almost all companies today. There are many benefits that stem from the increased use of intranets and Web-based applications that support an organization, and it is often easy to forget them or take them for granted. If we look back 10 years or more, there were intense discussions about the value chain and large-scale projects in business process reengineering (BPR). Now, when e-commerce is less hyped, we look back to the value chain as a way to visualize core processes, because it still serves as a good model for describing an enterprise. The purpose today should not be to implement BPR, but the value chain is nevertheless a good tool in ascertaining how mobile devices can enhance business processes. We will not describe how to use the value chain—most of you probably remember it. Instead, we have chosen five areas from different parts of the value chain where there is large potential for many companies.

The five areas are:

- Field service
- Sales force automation
- Telemetry—remote monitoring
- Transportation
- Dispatching and routing

In these areas, identifying tangible benefits will lay the foundation for solid business cases with improvements in cost reduction and revenue generation. Business cases that are based on customer communication are usually difficult to calculate because they tend to identify qualitative benefits, which are harder to estimate.

Here are some benefits from internal efficiency gains that are easy to calculate:

- Receive detailed work orders remotely
- Spend less time reporting at the client site
- No need to register reports from mobile workers in back-office
- Direct access to inventory and lead times for ordering
- Direct access to product information, manuals, and technical specs

Back-office registration is an especially time-consuming task. The quality of the information will improve if registration can be carried out directly on-site.

Here are some benefits gained from customer communication that are more difficult to calculate:

- Facilitating access to product information
- Making it possible to place orders from a mobile device

- Improving service through mobile channels
- Providing location information to your retailer

These functionalities will generate sales. Thus, it will be necessary to first estimate the increased volume they would generate and then estimate the value of each purchase or retained customer. This is not an easy task.

Another difficulty is that, when dealing with customer communication, there is no control over which mobile device each customer will choose to use or how it will be configured. This means that the application will have to leverage the same service to many different devices, which in turn will complicate development and maintenance.

Some processes will be applicable to many businesses, for example, time reporting, planning work orders, and checking invoices. First, let's look at some of these processes.

Time reporting is a very time-consuming activity in many companies. One of the reasons for this is that such reports have to be filled out retroactively—that is, you have to try to remember what you did earlier that week. With mobile devices, you will be able to carry out your reporting as soon as you have completed a task. Always having your mobile device at hand will make this possible. You will be able to report each assignment as soon as you've completed it or include all the day's assignments in a report that you send in at the end of a day. Time reporting is treated differently, depending on which line of business you are in. While some companies invoice the total time spent at a site, others only follow up time spent on internal procedures. However, most companies require their staff to report time spent on different tasks. Asking staff to spend 15 minutes every Friday to recall all the tasks they've performed that week is a waste of time.

Planning work orders—or any task planning for that matter—requires that you connect to a system to consider any new assignments. Being able to connect at any time and view additional tasks that have been registered since you last connected improves your efficiency. You will simultaneously be able to obtain any necessary client information you might need to perform these new tasks. If this functionality is combined with calendars and access to back-office information on customers, the efficiency of the mobile workforce can be increased significantly.

Checking invoices on a mobile device is another task that could be adopted by a number of different businesses. The time pressure placed on managers with invoicing responsibilities could be reduced if they were able to check all invoices from a mobile device. Today, managers have to check invoices after they have been made, after which the invoices should be sent to the client (preferably on the following working day). Most managers would consider it a blessing if they were able to take care of this process by way of a mobile device when out of the office. The technical solution could vary from organization to organization—invoices could either be sent to managers as messages or managers could connect directly to the ERP system and accept them online.

The opportunity to perform tasks directly and not have to make notes of them is, as we know, one of the greatest benefits of mobility. It increases enterprise efficiency in ways most people have never even thought of. From the technician who can find out if a spare part is available before heading to the next customer, to the sales rep who has 15 minutes to kill at the airport and takes the opportunity to call potential clients, mobile devices bring huge possibilities for increased efficiency.

Employees will be able to read their e-mail in a taxi on their way to the airport to check if something requires their immediate attention. We will be much less dependent on PCs, and this will increase the efficiency of people who work outside the office but who still want to be in control of what information is being sent to them. Efficiency can be many things—mobility is something that can empower an efficient work force.

Calculating Value Generated from Mobile Functionalities

Mobility brings value to enterprise processes, both to the mobile employee and to the internal processes that deal with administration. Many companies find it difficult to calculate these benefits. So, we need to find solid evidence—in the form of concrete figures—to obtain the necessary funding for projects.

Business value can be divided into two categories: quantitative and qualitative benefits. We need a structure that can easily provide calculations based on quantitative benefits; we also need to make the qualitative benefits as tangible as possible. As shown later in this chapter, field service automation gives rise to quantitative benefits that are relatively easy to measure, whereas sales force automation leads to more qualitative benefits. We normally find that several benefits stem from the implementation of any new functionality. Therefore, it is necessary to combine the value of all these benefits and implement them in a project.

The need for a solid business case is often underestimated—an example of this is when the employees involved in a project start to investigate where mobile functionalities can add value. When trying to obtain funding for a project, enthusiastic people within lower management find little acceptance from top management. The reason for this is unwillingness on the part of top management to invest, and this reluctance is often a direct result of the absence of a solid business case.

Quantitative Benefits

Quantitative benefits can be translated into figures that can be used when making calculations on a business case. This does not necessarily require that the benefits can be measured in dollars. They can, however, be translated into other concrete figures. For example, if a technician saves five minutes on every client call by having information available to him or her in a mobile device, a figure can be extrapolated that indicates the number of minutes a technician can save on every phone call. If you then multiply the number of calls made every year within the whole company, you will get a figure that can be used as a business case.

Here are some additional examples of quantitative benefits:

- **Cost reduction in back-office administration.** You can reduce staff in back-office.
- **Improve efficiency of technicians.** More tasks can be performed per day.
- **Reduce logistical costs.** Information on the fastest route between clients is always available.
- **Improved cash flow.** Lead time when invoicing will be reduced (calculation includes interest rates).

These are of course only a few examples. They are, however, probably the most common ones that are experienced when implementing solutions to improve internal efficiency.

Qualitative Benefits

Assigning figures to qualitative benefits is a more difficult task, because the benefits can be external (how the client experiences a service) or internal (how the provider improves a service). The reason it is difficult to describe client benefits is that the provider has to translate the benefits into increased sales and then calculate the value of each additional sale. Internally, qualitative benefits are often related to improved procedures, which in turn lead to fewer errors. So, to quantify qualitative benefits, it becomes necessary to calculate the value or cost of such issues as having incorrect input fed into the back-office system.

Here are some additional examples of qualitative benefits:

- **Better planning and utilization of employees.** The right person at the right place when you have information available on-site
- **Fewer errors.** Both the correction of errors and the negative impact if they aren't discovered, because the information is registered once and not on paper
- **Faster, more accurate communication and information.** Makes employees more efficient
- **More accurate inventory control.** Knowing what is available

Qualitative benefits can always be converted into figures; the difficulty lies in the parameters you use in the formula. Let's make the following calculation to illustrate the cost of inserting an incorrect value into a system.

Imagine a sales force consisting of 500 sales reps. Now, attribute a value to entering correct information into an order system. Assume that two orders are entered into the system too late or with incorrect information every week. Assume the cost of dealing with these delayed/incorrect orders (including loss in credibility vis-à-vis the client), is $50 per order.

500 sales reps × 46 working weeks per year × $50 per delayed/incorrect order = $1.15 million.

This kind of calculation might not be enough to justify a full-fledged investment, but if the argument is presented in a discussion about cost reductions in back-office administration, you will certainly strengthen your business case substantially.

Presenting the Benefits of Mobile Technologies

A convincing presentation of a business case is important when attempting to persuade top management of the benefits that are tied to a project. The importance of this becomes even more apparent when dealing with mobile projects, because this is a new area for many organizations. Here, we present a way of communicating a simple calculation in which the assumptions can be changed in the spreadsheet in the course of a presentation.

You do not need a complicated calculation. The simplicity in Figure 10.1 makes it very easy to communicate and discuss with others to ensure that the calculation is correct.

Value Proposition: Less Time Spent on Paperwork

VALUE PROPOSITION:	Less Time Spent on Paperwork	
MOBILITY FUNCTIONS:	1. Process invoices, billing, and get signatures 2. Mobile office functions: e.g., expenses, time sheet	
COST SAVINGS: #1		Example
	# of orders per day Average cost/hour per field worker Time saved by inputting info.onsite (e.g., service requests, orders, contacts) (minutes)	5,000 $40 10
	Subtotal #1	$7,666,667
#2	# of field workers Time saved in administrative tasks/day (e.g., time sheet, expenses, client info) (minutes)	1000 10
	Subtotal #2	$1,533,333
	Total Cost Savings per Year	$9,200,000
REVENUE GENERATION:	Extra service calls/week Average revenue per service call	1 $200
	Total Extra Revenue per Year	$10,000,000

Figure 10.1 Presenting the benefits of mobile technologies.

Revenue generation is the preferred model in some organizations. There has, however, been a strong push for cost reduction in recent years, which implies that business cases should focus on that instead. Naturally, you will get cost savings or revenue generation, but not at the same time.

Field Service

First, let's define what we mean by field service:

> *Field service refers to all business processes that support the use of a product or service on-site.*

The reason for not addressing this subject together with ways of improving the sales force is that many characteristics differ. For technicians, the specifications concerning what parts are included and the history of the product or service need to be available when they are at a client site, for example.

Almost all employees working with field service at client locations are short on information about what they are going to repair. This is the case for anything from fixing a stove in a house to repairing assembly-line equipment in a plant. In the future, we will be able to support unanticipated information needs by making information accessible from mobile devices. We will also be able to access dynamic information such as inventory levels in real-time.

Just imagine the benefits that can be gained if a technician has access to the information that most companies have stored in their internal system (their intranet). If some spare parts are needed for the next job, technicians can simply order them from their mobile device. That way, a task can be completed when they leave a client location.

An increasing number of companies are connecting their production units directly to IT systems. This enables them to monitor processes in real-time, which increases efficiency because they know if something is going wrong or if spare parts are needed. In this way, technicians can have direct access to the same information while standing next to the equipment; they will be connected wirelessly to both the machine in question and the IT systems in the back-office. Providing technicians with access to manuals, the spare-parts inventory, and documentation of recent repairs carried out by other people will definitely increase their efficiency. When discussing field service, we normally imply mobile employees outside the plant, but we will see that many benefits come from employees connected wirelessly inside the plant as well.

Let's look at the standard field service process:

Identify → Plan → Perform → Report → Follow up

First, there has to be a service need, which is normally handled by some kind of dispatching function. The information is registered in the administrative system, and then the dispatcher assigns someone to the task.

- *Identify work* means that field technicians check the system for new assignments at regular intervals or they are given some kind of alert (potentially an SMS) to bring their attention to a new assignment.
- *Plan work* is the normal procedure of accepting an assignment, deciding when the work will be carried out, and knowing that the appropriate equipment will be available for the anticipated tasks.
- *Perform work* could include using check lists, ordering spare parts, and accessing knowledge bases for reuse of similar solutions to a specific problem.
- *Report work* includes registering what was wrong and how the problem was solved, if the task was specific in nature. This also refers to adding information such as how much time was spent on the task.

- *Follow-up work* is an interesting topic because the client would normally like feedback on what happened and how it can be avoided in the future. This also provides the opportunity to get a signature from the client that he or she accepts the work done. This simplifies invoicing procedures. This also provides an opportunity to send a conformation to the client, for example via e-mail, with a copy of the work order. Invoicing can now actually be done while the technician is still at the client site, which significantly improves cash flow.

Business cases for field service are relatively easy to calculate. The greatest benefits do not, however, emerge where most people expect them to. When a technician reports at a client site, he or she improves information quality and saves time. This time can now be spent on another customer assignment. The entire administration process that usually takes place after the technician has left the client is reduced by 50–90 percent. Today, the technician's documentation of the task that has been carried out is left to someone who is supposed to register all the information into the system. The fact that someone else is interpreting the technician's information sometimes leads to misinterpretations. This whole stage in which staff administers information can be reduced by at least half. In parallel, we can calculate the value of the technician being more efficient by having access to spare-parts inventory and knowledge bases containing solutions to similar problems. These are qualitative benefits, however, and it is often difficult to create calculations based on them. On the other hand, reduction by 50 percent of a back-office task is quantifiable and can much more easily be translated into saved dollars. Figure 10.2 describes different functions that should be taken into consideration when building the business case.

Sales Force Automation

When sales representatives meet clients, they should read the notes from the last meeting before entering the room. This seldom happens because the sales executives have a PC-based support tool, which is not available outside the office. Thus, sales representatives currently have too little support from the internal systems, and this has a direct negative impact on sales results. Now, imagine that the reps have access to the notes from their last meeting in a mobile device. It will take them one minute to read through them before the meeting—they might even do it before getting out of the car in the client's parking lot. The device will be connected to the back-office systems, so the notes taken during the meeting will automatically be entered in the sales support system. That way, all the information from the meeting will be available in the PC when the reps return to the office to write a quote.

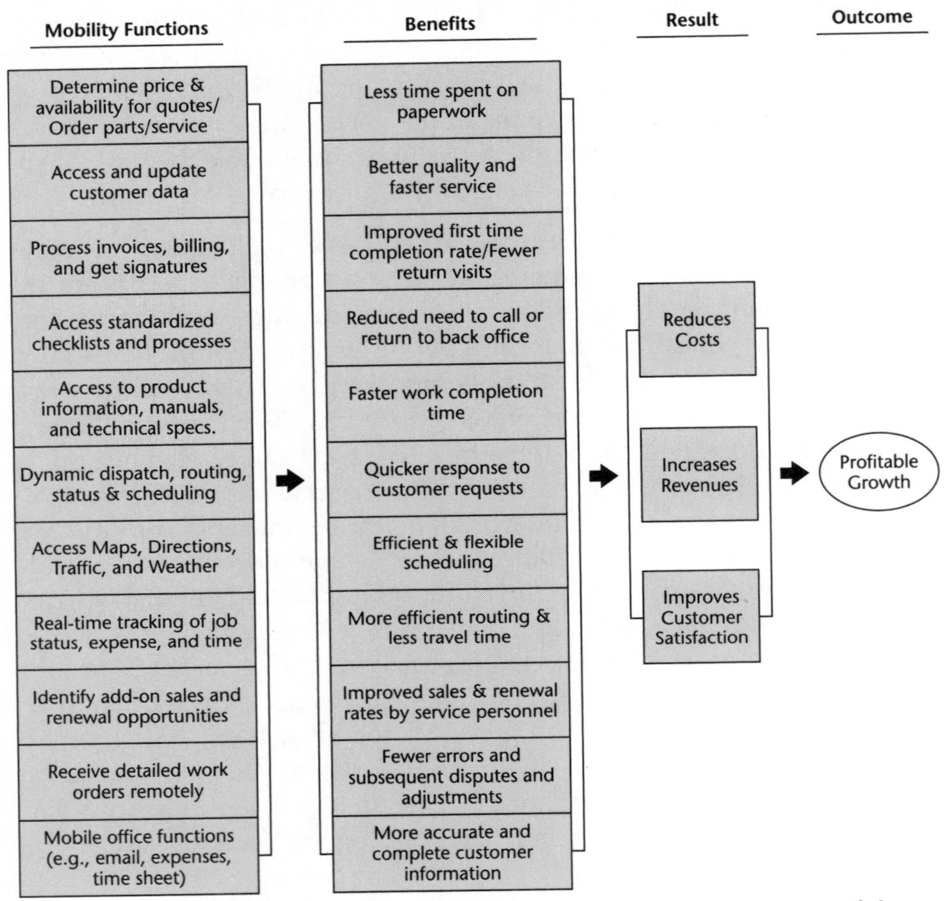

Figure 10.2 Field service mobility reduces costs, increases revenues, and improves customer satisfaction.

Here are some more examples of how the sales force can benefit from mobile devices. They are generic to ensure that most companies can translate them into their own situation and make quick cost/benefit calculations that will justify investments.

- Sales reps will be able to see how their colleagues are selling to similar customers, and this can give them ideas on how to generate new business.
- Sales reps will have instant access to information concerning the supply of the products they are selling. Many businesses work in a dynamic environment with rapidly shifting stock levels. If some

products are out of stock, the reps will be able to recommend another product. Furthermore, pricing will be set centrally, so the reps won't have to make difficult pricing decisions. All they have to do is check their mobile device when negotiating prices with customers.

Now, let's look at the normal sales call process:

Plan meetings → Meet client → Administer meeting

- *Planning meetings* doesn't only mean deciding and booking the appointment. It should also include a preparation activity for the meeting that goes through several stages. The very last step in the preparation process should be reviewing notes about the discussion that took place in the previous client meeting. In an ideal world, the preparation process should be carried out in a structured way and well in advance of the meeting. Unfortunately, too many sales reps travel around without the time to sit down and do this prep work in an office environment. Therefore, they make their preparations in a hotel or at home without access to all relevant information.
- *Meeting the client* with all relevant information at hand. Most companies currently sell relatively complex solutions or sets of components to their customers. There is usually a large amount of information that is impossible to remember by heart—even when the products or services are packaged, certain parts might have lead times for delivery and prices can fluctuate frequently. The key application will be to have the support from the back-office systems extended to the location where you meet the client. This is indeed an invaluable service.
- *Administering meetings* is something sales reps often have a guilty conscience about. Administration is a pain for most sales reps. Everyone knows it has to be done, but it is often put off to a later date. By then, however, they have forgotten some relevant information that should have entered into the back-office systems. With the mobile device, admin tasks can be carried out at once, and an opportunity list can be filled out together with the client. At the same time, reps will be able to take notes on the products/services the client would like to have information about. This, in turn, will reduce the amount of administration that needs to be carried out when back at the office.

As with other opportunities based on mobile solutions, there is no single feature that in itself will generate a huge amount of added value to an enterprise. It is rather the combination of all new mobile features that will bring added value.

As we have seen, business cases for the sales force clearly differ from those for field service. Whereas field service mobility gives rise to quantifiable benefits that can be calculated quite easily, sales force mobility generates benefits that are more qualitative in nature, which means that calculations will be more difficult to express in earned/saved money. How do you establish the value of a client experiencing better quality delivered by a sales rep? Will the client buy, on average, 5 percent more? If reps can be more specific concerning delivery times and prices when sitting down with the client, will that cause sales to go up by 10 percent? Everyone understands the value of the representative selling 5 percent more, but this figure is very much a guess. Are you aware of the real figure for your company? Figure 10.3 describes the different functions that should be taken into consideration when building the business case.

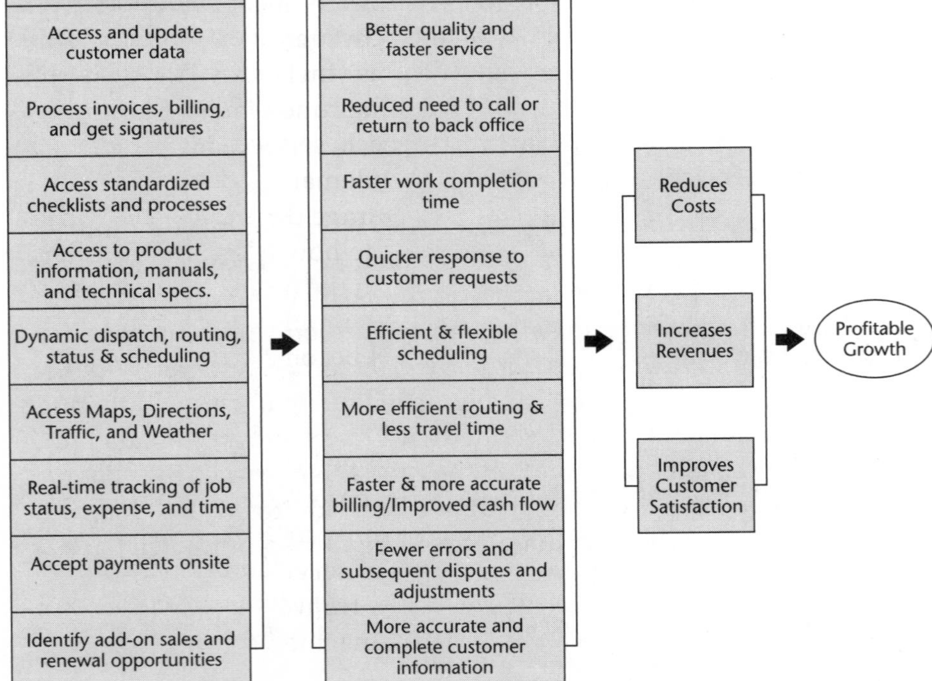

Figure 10.3 Consider these issues when building your business case.

Telemetry: Remote Monitoring

Remote monitoring: Connecting equipment to remote surveillance via a network of some kind.

Remote monitoring is currently being utilized internally in many companies. They are connecting process lines to process surveillance, thus making it possible to monitor the manufacturing process from computers located inside the company. This enables the automation of emergency alerts. One example of this is messages that are sent in SMS form to a cell phone in case of an emergency. Another example is when the system is connected to a beeper that sends out a signal as soon as a problem is encountered. In this way, staff knows that their attention is required, and they can monitor the equipment from a distance. They then have the possibility of using a standard Web interface to connect to the system from a remote place. Sometimes, the necessary adjustments can be carried out remotely. In other cases, a technician is needed to repair the equipment.

In the future, we will have the opportunity to connect to machines outside the company; these might be copiers, trucks, cooled containers, process equipment, or coffee machines. It will be possible to connect to virtually any information available in the machine. You will first be alerted to the problem, and you will then be able to start analyzing the problem from a distance. Planned maintenance will still exist, but even that will be possible to perform remotely. The greatest advantage, however, will come from the possibility of connecting to the system as soon as you have received an alert.

In addition to the alert and remote monitoring functions, mobile devices can be used to determine the frequency of vehicle service, which is an area where costs can be substantially reduced. This demands that you know the mileage of the truck, or how many times the garage door has been opened. Companies use advanced calculations to decide how often service should be performed, but it is sometimes hard to obtain the necessary information when deciding at what point in time a vehicle should be serviced.

Refill frequency is yet another area that will become more efficient as a result of mobile devices being installed in vending machines—soda or coffee machines are two simple examples of this. If you can reduce the number of times you refill the machine by 20 percent, which is realistic, you will instantly save 20 percent of the refill costs. This can be achieved with the help of simple technology. The main problem is determining when existing machines should be taken out of service. The new technology usually has to be built into the machines from the outset (it is very expensive to add this functionality to existing machines).

Remote monitoring is frequently used in the transportation industry and is, of course, highly dependent on what kind of equipment we are discussing. A truck, for example, will send an alert when the oil temperature is too high. You will also be able to get information on oil levels, transmission status, mileage, and run times. It will even be possible to download new software to the truck that will enhance its performance without having to take it to the mechanic.

If you are driving your car and you experience engine failure, your insurance company will immediately be made aware of the problem and a mechanic will be dispatched along with a substitute vehicle. Picture it—the insurance company will be calling you before you call them. Now that's improved service. In fact, this scenario isn't too far away. There are already cars that automatically connect and send a signal to a surveillance system as soon as an airbag has expanded, thereby enabling localization of the car. Emergency service centers are now testing a voice communication service that will make it possible to call drivers and ask them what has happened.

Transportation

The entire transport industry is in the process of entering mobile ventures, and the incorporation of mobile devices in vehicles is increasing. When transportation companies plan which trucks will transport what goods, the information needed will depend on the situation (knowing the location of the truck, where it has been, which route it will take, and how many goods it can carry). Information about how vehicles move between customers can be obtained either through messaging or actual viewing on a screen by way of mobile communication.

Fleet management systems currently send very small quantities of data when using existing mobile systems, which are mainly delivered by Mobitex. However, the need for drivers to get information and guidance is increasing and this demands another kind of technical platform, that is, one that will provide a graphical user interface with high bandwidth, such as 100–200 Kbps. This is, in fact, currently being delivered by niche suppliers in the automotive sector and is built on GPRS/3G.

Specialized equipment is currently being used to equip taxi drivers with maps that provide them with information on their current position and the fastest route to an address. In the future, however, taxi drivers will also be able to access content through an internal system that can predict where transportation needs will arise. This will increase efficiency in the sense that they will use standard services besides services that have been specially developed for taxi organizations.

Travel support systems are another useful mobile application; they will help you plan your trip from the office. The application will continuously update changes in departures and synchronize them your client meetings through the Internet. You will no longer have to rush to the airport only to find that there is a two-hour delay. Your mobile device will keep you connected to the travel-planning system even when you're out of the office. The travel-planning system will also be connected to your expense report, which you will be able to constantly update during your trip so that all costs are accounted for when you are back at the office.

Dispatching and Routing

Keeping track of moving equipment (such as trucks) and staff outside the company is becoming increasingly important. This is already quite common in the transportation industry, where it's possible to know where a specific truck is located at any time as well as how much free space it has available. Needless to say, this increases efficiency considerably. Similarly, sending a taxi to an address is a rather simple task—the trick is to know which taxi should be sent. As we have seen, GPS systems are already in place that keep track of taxis and trucks with the aim of improving efficiency. In the future, companies will start using this functionality to monitor their mobile work force.

Let's say the central heating system has broken down in your home and you call the round-the-clock service for a technician. You will probably have to wait a couple of hours before he or she arrives. Today, companies have to call all their technicians and ask for their availability, which is a time-consuming task. In the future, however, companies will know where their technicians are. They will also know if they can take on an urgent assignment. In this case, sophisticated positioning systems aren't necessary. Base-station positioning will suffice to provide information on the location of technicians within a radius of approximately 200 yards. What's more, if they are equipped with devices that enable them to accept assignments and report back to the office when they've been completed, deciding what technician should be sent to fix the central heating system becomes an easy task.

Many companies that provide this kind of service are, of course, rather small. They probably know where their colleagues are, and all they have to do is call the person they think can take on a job. And they often make the right choice. This is the reason why some services are often provided by small local companies—local technicians can take a break in an extended job to make an urgent call elsewhere. Large companies with connected employees will, however, have a competitive edge compared to smaller companies when it comes to providing clients with fast service.

Summary

This chapter covered how different sectors use mobile devices. There are obviously many applications that are specific for one or two sectors. However, most applications are generic in the sense that they apply to many industries and normally create an extension from an intranet to the mobile worker. This provides the same information on-site as users have at the office.

The conclusion is to start with a simplistic approach. Determine what kinds of information are available in the office and what approach would increase the efficiency of the mobile worker.

CHAPTER

11

Microsoft and CGE&Y Case Studies

Adam Kornak
Jorn Teutloff

Introduction

To provide you with a few ideas that will get your own thinking started, this chapter presents a few real-life case studies of mobility and wireless technologies at work. These case studies reflect previous projects that project teams at Microsoft and CGE&Y performed for select clients in various industries. To maintain client confidentiality, their names have been masked and strategic details have been omitted. Yet, we hope you will find these case studies useful in that they illustrate mobility principles applied in the real world by pioneers who realized that the only way to derive value from new technologies is to learn about them, pilot a few initiatives, assess the results, and then selectively invest in the rollout of the initiatives that hold the most promise.

All case studies presented on the following pages were created over the past two years; some of the projects have evolved from limited deployments to company-wide implementations, others have been trenched back—all are continuously evolving as mobile technologies twist and turn on their way to becoming deeply engrained in the way companies operate.

Although the case study examples are presented here at a fairly high level, several are discussed in great detail in the companion book to this text. The *Cap Gemini Ernst & Young Guide To Wireless Enterprise Application*

Architecture (Wiley Publishing, Inc. 2001) illustrates at great lengths the conceptual business architecture, information architecture, information systems architecture, and technical infrastructure architecture of four of the case studies presented here. Especially if you are a systems architect, you might want to review the companion book for additional information regarding the development of mobile enterprise applications.

Wireless Field Service Automation

The first case study describes a field service solution. Let's assume a client by the name of Move, Inc. The company manufactures heavy equipment used to move parts within a customer's plant. Move, Inc. manufactures, sells, and services this equipment, yet the main source of profit in the company's business is after-sales revenues in the areas of service and spare parts (the sale of spare parts has higher margins than the sale of the original equipment). Move, Inc. has decided to invest in the use of mobile devices, and the first step in their mobile venture is to supply their mobile technicians with handheld devices. Pocket PCs are the mobile devices chosen to enable the technicians to gain access to back-office systems and report directly from client sites. Move, Inc. employs more than 1,000 technicians throughout Europe whose main task is equipment service and repair. The technicians use light trucks to drive from client to client, carrying frequently used spare parts as well as the necessary equipment to perform the service. Move, Inc. sometimes has more than one parts depot per country, depending on the number of technicians and the distances to different customers. In addition to being spare-part warehouses, the depots provide the technicians with administrative support.

Today, the administrative staff at Move, Inc. deals with more than one million work orders a year. There are at least four copies of each order: one for the client, one for the technician, one for registration purposes, and one for archiving purposes. The solutions brought about by deploying mobile technologies will eliminate the need for numerous paper versions of each document. Thus, back-office staff will be able to significantly reduce the amount of time spent on data entry, registration, and error tracking. Furthermore, the number of errors will be reduced because the person performing registration activities will have access to all pertinent information in electronic format. Today, the back-office staff has to interpret what the technicians have written by hand, which frequently leads to mistakes.

The Pocket PCs to be distributed to the field force come with software that enables the technicians to work with the device without having to connect to the back-office system. The synchronization functionality will enable the exchange of updates whenever appropriate. Later, we will see that this will be done on a frequent basis.

At the start of the project, the aim was to reduce administration costs and increase service revenue through better utilization of the technicians and streamlined data capture. As we will see later, there were a number of additional parameters that were also improved, including customer satisfaction and logistics.

The Solution

The mobile technology that Move, Inc. developed for the company's technicians is a *thick client,* meaning a system that entails both the application and the database residing in the handheld device. The database is synchronized with the middleware, which is directly connected to the ERP (enterprise resource planning) system. Thick clients have two advantages compared to thin clients: speed and the ability to always work without a wireless connection. With thin client solutions, on the other hand, the only functionality residing in the device is a browser, which demands wireless access at all times. For thick clients, synchronization takes less than one minute and can be performed whenever the technician needs information or wants to send registered data. The reason for using middleware to handle synchronization is simple—this functionality does not exist in today's ERP systems. It is apparent, however, that most ERP vendors are ensuring that their next generation of applications will have this functionality as an optional component. The same applies to client applications, which are also optional components (that is, the application residing in the PDA).

The importance of a good user interface must not be underestimated. Most technicians are not familiar with operating a PDA. For this reason and to justify changing from pen and paper to operating mobile devices, it is of the utmost importance to provide the end-user with an application that is both user-friendly and efficient. There are two main issues to keep in mind here: how data will be captured and how the presentation of the information will be adapted to the process.

When capturing data or information with the mobile device, the goal should be to never have to use the device keyboard. Wielding a stylus to tap in character after character simply isn't efficient, even if an external keyboard is attached to the unit. Instead, it should be possible to perform all input by choosing options from drop-down lists in the application. Although it might be difficult to foresee all possible data options when designing the application, it is important that situations demanding the use of a keyboard be held to a minimum.

For this reason, it is critical to involve the end-users when designing and building the application, because their involvement will ensure the development of a product that truly supports them and improves their efficiency. It would be myopic to attempt to build a solid application with easy-to-use functionality without getting feedback from the people who will have to work with the application on a daily basis. Consequently, simply converting the paper-based form that the company's administrative staff used for many years would have resulted in the loss of many of the unique benefits that a mobile solution affords.

The Process

In order to support technicians and help them perform their daily tasks more efficiently, the application needs to be based on work processes. Creating work processes that meet the needs of different company cultures is, however, a great challenge. In this case study, the issue was how to create a single work process that meets the requirements of staff working in different regions, such as Italy, Portugal, Germany, and Sweden. In general, it is more difficult to create a new workflow for an existing organization than to develop one from scratch.

Move, Inc. is working with a traditional process containing the following steps:

- Identify work
- Plan work
- Perform work
- Report work
- Follow up on work and create invoice

These generic steps are frequently followed by organizations when identifying organizational workflows to be adapted for mobile technologies. There are surprisingly few differences between organizations in different sectors, be it telecom or life science. The need to access information and report findings on-site makes the solutions very similar. The following sections describe each step, focusing on how mobility changes the way that work is performed.

Identify Work

To begin with, it is necessary to perform an analysis of service needs at the client site. Although most service calls are planned, a considerable number of client calls will have to be attended to at short notice (equipment breaks down and needs to be fixed as soon as possible).

There are three ways of creating a work order:

- **Planned service.** The work order is automatically created based on the type of equipment and service intervals that have been agreed to with the customer.
- **A client calls the dispatcher.** The dispatcher then assigns a technician to the job along with a work order, which is simultaneously created in the back-office application. The work order is then synchronized to the device. If it is an urgent call, the application will automatically send an SMS to alert the technician and ask him or her to synchronize, after which the technician will obtain all relevant information about the work to be performed.
- **On-site need.** The technician observes a need at a client site, or has been asked to look at some other equipment that isn't registered in the work order. The technician then registers all relevant information directly into the device. The information is later synchronized in the back-office system.

The third way of creating a work order is enabled and facilitated by the new functionality. Today, the technician starts out by mending the equipment and then proceeding to call the back-office, asking them to create a work order, which he or she fills out later. The problem is that technicians often forget to fill out the work order and as a result, the client never has to pay for the work performed. You might argue that this is good service, but it is not a good business practice. When clients receive speedy service by technicians who are already on site, they don't mind paying for the additional service performed.

Now, let's look at the process from the technician's point of view. Here, we need to find a way to make the technician recognize the value of using the device. Too many projects have failed due to the discrepancy between theory and practice.

Plan Work

After synchronizing the PDA with the home base, Move, Inc.'s technicians have all the relevant information they need to perform a task. All that remains is to decide in which order they want to take the calls. Today, there is no route-planning support available to technicians. Instead, they simply plan their weeks based on parameters such as where to spend the night and how to reduce travel time. This behavior will eventually change with the increased use of mobile functionalities.

When all work orders are managed in electronic format, assignments can be prioritized according to urgency, client location, traffic conditions, and other relevant information. In addition, routing can dynamically be changed according to new information, such as a customer canceling an appointment at the last minute.

Another impact of mobile technologies on planning the work involves the truck's spare part inventory. Using the mobile device and the work orders it contains, technicians will be able to quickly determine which spare parts a job demands, and verify that such parts are available in their truck or if they need to be collected before leaving the depot.

Perform Work

While performing a service task, technicians will use mobile devices mainly to review and verify service histories, service eligibility, and spare parts, and to view checklists to ensure that all activities have been carried out according to the work order. While on the job site, technicians will also be able to order spare parts so that they are available and stocked in the truck when technicians perform the next service at the particular customer site. Conversely, every spare part used at a customer location will be registered in the mobile device to ensure that the truck is replenished the next time it arrives at the depot. Today, paperwork is used to order replenishments, and technicians get a box with spare parts when they arrive at the depot. In the future, the process will be much simpler. Technicians will merely have to synchronize their PDAs once they've completed a task, and the parts box will contain all replacements, including the parts that were used at the last client call.

Report Work

As mentioned earlier, the quality of reporting is largely dependent upon how far from the source the reporting is entered into the back-office system. Enabling technicians to place reports directly into the system when on-site will substantially improve reporting quality.

There are several aspects in the reporting process that will be rendered more precise thanks to mobile devices. First, although time consumption is not of great interest to invoicing procedures (most service agreements aren't based on hourly rates), this kind of information is valuable when conducting internal analysis of completed assignments.

Second, incorrect registration of the problems that caused the customer to call a technician can be greatly reduced by checking off the prepared options stored in the PDA—what's more, this can be done when standing next to the equipment. The most important issue is the reduction of free-text input. The entire registration procedure should therefore be based on a "drop-list" application whereby technicians simply choose between different predetermined options. This new functionality enabled by the mobile device supports and guides the problem analysis performed by technicians, taking them through a number of steps in less than 15 seconds.

Follow up on Work and Create Invoice

One of Move, Inc.'s goals is to more closely involve the client when performing service calls. The intent is to improve the level of service and professionalism. The company can achieve this by using the mobile device to show work orders, checklists, and error statistics directly to the customer's contact person. When the two parties are in agreement, the client will be asked to sign the work order directly on the device.

When the client has signed the work order and the data has been synchronized, an invoice can immediately be sent from the back-office system to the customer. In addition, the customer receives an electronic copy of the work order, including the signature of the person who signed off on the work performed. Thus, the manual process that used to take weeks can be carried out while the technician is still at a client's site!

What's the Business Value?

A discussion of the business value of mobile solutions needs to be divided in two separate parts, *quantitative benefits* and *qualitative benefits*. The difference between the two is that it is easy to calculate quantitative benefits, whereas the qualitative ones are difficult—sometimes even impossible—to express in dollars. The purpose of this discussion is not to present any detailed figures or calculations. Instead, we will provide you with an illustration of how a business case can be built. Let's start with quantitative benefits, where the most tangible improvement comes in the form of cost reductions in back-office administration.

Quantitative benefits can be measured and translated into dollars, but you might be forced to translate the figures to receive an amount. For example, if it can be demonstrated that back-office administration costs will be reduced by 50 percent, you will have to calculate the total costs connected to running the back-office department and then deduct half. This includes salaries, benefits, office supplies, and so forth. Be careful not to stop at merely calculating theoretical cost savings; if your analysis shows that due to the deployment of mobile technologies your admin staff is freed to the amount of, for example, 2,000 hours a year, but you do not eliminate a full-time position from your payroll, the theoretical savings might not translate into actual bottom-line numbers.

Qualitative benefits, on the other hand, are far more difficult to express in monetary terms, because they deal with such intangibles as quality of service and customer satisfaction. The following questions are difficult to answer in monetary terms:

- How much is good service worth?
- How much is more accurate inventory control worth?

The key issue here is to try to somehow translate the qualitative benefits into a concrete estimate, even if it is difficult to be entirely accurate. Here's an example: let's assume that a company is experiencing a problem where 1 invoice in 20 is incorrect. If this error rate could be cut in half, is it possible to express the savings in monetary terms? The answer is yes, if you can accurately calculate the cost of each invoice, which entails costing various activities such as preparing, processing, filing, reporting, and error correction.

Quantitative Benefits

When dealing with quantitative benefits, we need to identify how to translate them into dollars. Let's take the most important benefits that accrue to Move, Inc. and see how this can be done.

- **Reduce costs of back-office administration.** Because the reporting of equipment issues and performed work will be done at the client site, back-office staff will no longer have to deal with this task. Some administration of the workforce will remain, but the time-consuming and error-prone tasks of interpreting what the technicians have written and entering reports in the back-office system will be eliminated. The administrative cost associated with transcribing a work order can be eliminated.

  Benefit. Multiply the administrative cost per work order by the number of work orders.

- **Improve the efficiency of technicians.** Technicians can perform the entire planning process in one step, which means that they no longer need to gather client-related papers to find specific equipment checklists that describe how the work is to be performed. Administrating the reporting process and bringing the necessary papers to a client is time-consuming. In addition, technicians won't need to call someone connected to the back-office system to check the availability of spare parts—all that information will be available in the mobile device, which further saves time.

  Benefit. Multiply the number of minutes saved on each work order by the cost per minute and the number of work orders.

- **Improve cash flow.** Instead of the two weeks it currently takes to complete the work order registration process, it will be possible to send invoices while the technician is still at the client site. Today, technicians normally return to the depot and hand in their paperwork less than twice a week, after which the administration needs to transcribe the work order, create an invoice, and mail it.

  Benefit. Multiply the reduced need for working capital by your interest rate.

  Reduce logistical costs. Lower inventory levels will be achieved as a direct result of the increased control of consumption. Indeed, by using mobile technologies it will be possible to receive updates regarding used spare parts on an hourly basis. The manual paper-based process of today requires Move, Inc. to keep large inventories of spare parts, because the company does not know for days—sometimes weeks—which parts have been used at a customer site.

  Benefit. Multiply the reduced level of inventory by the interest rate.

Qualitative Benefits

When looking at qualitative benefits, we need to relate them to values of some kind, as described earlier. In the case of Move, Inc., some of the "soft" or strategic values (better service and more accurate inventory) support the company's professional image and brand positioning. Qualitative benefits are usually offshoots of solid business cases that present quantitative, measurable benefits. Few organizations invest in mobile solutions without a solid business case. Once they have invested, however, the intangible value of qualitative benefits contributes to the project's success.

Move, Inc. has experienced several qualitative benefits in spite of the fact that the business case was built on measurable benefits. Here are some qualitative benefits Move, Inc. identified:

- Improved invoicing accuracy
- Faster and more accurate communication and information access
- More accurate inventory control
- Faster replenishment
- Higher competitiveness, stronger corporate image, enhanced staff motivation
- Better planning and utilization of service technicians
- More efficient "startup" of new engineers and back-office staff

When combined, both qualitative and quantitative benefits provided enough of a business case for Move, Inc. to develop and deploy the mobile solution we described in this case study. This high-level study illustrated some of the benefits that can accrue to companies that support a large number of mobile employees, whether they are service technicians, sales representatives, or mobile executives.

The next study examines the application of mobile technologies in a distribution environment. Please note that the Wireless Commodity Exchange, Wireless Sales Force Automation, and Mobile Information Dashboard case studies are discussed in even further detail in the *Cap Gemini Ernst & Young Guide To Wireless Enterprise Application Architecture* (Wiley Publishing, Inc. 2001).

Wireless Commodity Exchange

In this case study, we will discuss a wireless solution deployed for the operator of a metals exchange. The Internet-based commodity exchange functions as a clearinghouse between buyers and sellers of metals products. Our client had effectively deployed this exchange via the Internet to create a virtual market space that benefits both buyers and sellers by eliminating the time-intensive and costly tasks associated with locating each other in the physical world.

The main benefits that accrued to the buyers who subscribed to this Web-based exchange included the following:

- Obtaining a single point of contact with all metal suppliers
- The ability to streamline contract negotiations

- Approvals and management
- Sharing demand and production planning information with suppliers
- ERP integration to facilitate global inventory reduction
- Establishing a secure, efficient and cost-effective way to engage in repeat order processing for ongoing contract buying

All of these benefits effectively allow buyers to move towards near real-time buying cycles.

Subscribing sellers, on the other hand, benefited from a single point of contact with a larger group of existing and potential buyers, realizing improvements in the speed, cost, and accuracy of order entry and fulfillment for repeat customers, reducing the risk of stock-outs, lowering freight rates as a result of more efficient use of transportation, and integrating ERP systems to facilitate global inventory reduction.

As you can see, the exchange, a neutral Web operation that neither represented a specific buyer or seller, not only had as its goal the matching of two parties, but also aimed at significantly streamlining the actual transaction process that traditionally has been marred by inefficiencies surrounding the manual interchange of information between buyers and sellers. Because the exchange was electronic in nature, it offered the opportunity to connect both the buyer and seller's back-end information systems for product requirements matching, payment data transmissions, and shipment routing and tracking data exchange. Thus, the exchange held the promise of significantly enhancing a complex supply chain in an effort to accelerate transactions, maximizing customer service and asset utilization while minimizing total delivery cost and inventory investments.

The supply chain function of the metals exchange is characterized by several critical features, including speed, connectivity, collaboration, optimization, and visibility. The Web-based electronic exchange offered fast processing, many-to-many connectivity, collaborative information sharing, and real-time visibility into product availability and order and shipment status. It is the last feature, visibility into the supply chain, which was to be wirelessly enabled. We are going to explore the wireless solution—its scope, functionality, and benefits—in a little more detail in this section. Again, for detail surrounding information, information systems, and technical architecture for this case study please see the previous book in the series of Cap Gemini Ernst & Young mobile technology guides.

Conceptual Business Scenario

As previously mentioned, our client, the operator of the electronic commodity exchange, was interested in making the system's visibility feature available on a mobile device, in this particular case a Pocket PC handheld. The goal was to allow mobile users to monitor and control customer fulfillment networks by tracking order flows through the transportation exchange network in real-time. Figure 11.1 illustrates the wireless visibility application by showing its data sources and its wirelessly enabled functionality components.

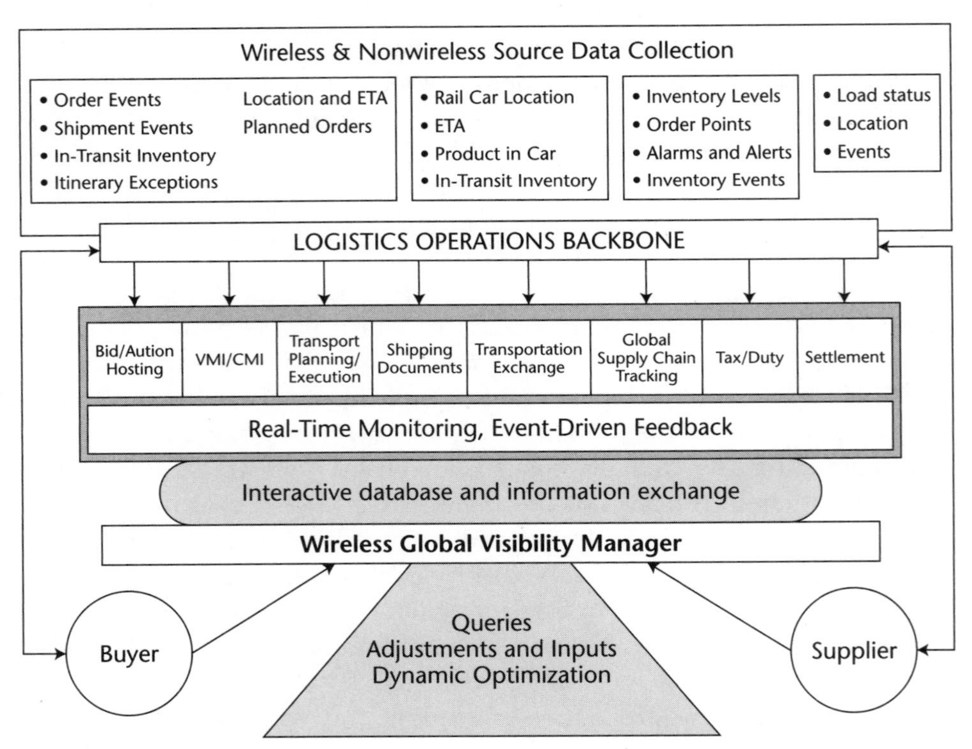

Source: CGE&Y Analysis

Figure 11.1 Wireless visibility application—data sources and wirelessly enabled functionality components.

Specifically, the wireless application would allow its users to monitor, view, and manage several aspects of a transaction in progress, while being able to communicate with other members of the system. The set of features includes the following:

- Order/bidding management functionality allows for the creation of a bid or an order, the modification of same, or its cancellation from a mobile device.
- Vendor managed inventory (VMI) services provide the ability for a supplier to constantly monitor a customer's inventory and trigger automatic replenishments should levels fall below a specified level.
- Transportation planning and execution entails the planning, optimizing, and directing of shipments from a seller to a buyer.
- Shipping documents is a component that assists with the reconciliation of bills of lading and invoices.
- Transportation exchange provides for the effective matching of shippers with independent carriers.
- Global supply chain tracking provides the capability for suppliers and buyers to track a shipment throughout the order fulfillment process. In addition, this feature issues real-time exception-based actionable alerts should certain parameters, such as the estimated time of arrival at the buyer's warehouse, be out of range due for a delay in the customs clearing process, for example.
- Tax/duty functionality assists the parties to the transaction with preparing all required documentation throughout the process.
- Last, but not least, the settlement module assists with managing the payment process.

By providing insights into the distribution mechanism, the wireless visibility application allowed users on the buyer and seller sides to monitor supply chain activities especially surrounding the tracking and tracing of shipments, to receive alerts when delays occurred, and to quickly assess the progress of an order as it made its way from the seller to the buyer. In addition, the application provided value to the organization and its stakeholders by allowing for communication with other users via voice and data messaging.

Let's look at a use case that illustrates how the application performs. In our example, a buyer of sheet metal is running low on inventory and realizes that the contract supplier won't be able to meet the company's demand on short notice. The buyer must now seek to replenish its inventory from other suppliers—and thus logs onto the exchange to quickly connect with several sellers who may carry the grade of product required. Taking advantage of the electronic market space, the buyer puts out a bid for sellers to reply to.

Now, many sellers, hoping to connect with buyers for their products, monitored the exchange. One of these suppliers was subscribing to the wireless application we are describing here. The application allowed that company's sales representatives to maintain connectivity with the exchange from any place, at any time, while the company's drivers used the application to provide real-time data on their positions, their truck's inventory, and other details surrounding the shipment such as expected times of arrival.

As soon as the company looking for sheet metal placed its bid on the exchange, a field sales representative who was geographically close to the buyer used his Pocket PC device to respond, having real-time visibility into his company's inventory levels and spot pricing levels. Not only was this representative able to immediately reply with specific information regarding inventory availability at the company's warehouses, but he also was able to search his company's expected deliveries. This lookup revealed that the rep's company was about to receive a shipment of the type and grade of sheet metal the prospective buyer was looking for. The rep immediately cleared the paperwork—using his wireless device—and received approval to sell the inventory still on the truck, while the driver was rerouted on the fly to drop the shipment at the buyer who placed the bid instead of at the company's warehouse. The buyer was able to replenish his inventory on short notice, and the sales representative using the wireless application was able to make to book an additional sale and establish a relationship with a new buyer.

Having witnessed the wireless application in action and the empowerment it bestowed on the buyers and sellers who subscribed to the service, let's look at some of the technical aspects required to make it all happen.

Defining the Solution

One of the first things to realize when defining the solution for an application of this magnitude and complexity is to use an iterative model for design. In other words, don't try to use the "big bang" approach and build the entire application at one time. When thinking about designing an application, it's generally always easier to think of the build process as unique layers or components that combine to form the overall solution.

The focus of this section is to understand how the wireless exchange application is technically designed at a high level. The objective of these next few pages is not to list each API (application program interface), or each connection to each component to the front-end and back-end interfaces. Rather, it's to help you understand how the business layers transition into a technical design that can ultimately be used to begin building the application from the ground up. In fact, it would likely take more chapters than we have available in this book to detail all the various triggers, APIs, services, and layers of the application. Let's start by illustrating the high-level architecture of the application that was used created in the early phases of the wireless exchange project. Figure 11.2 shows a layered architecture and the components and their respective connectors used in building the application.

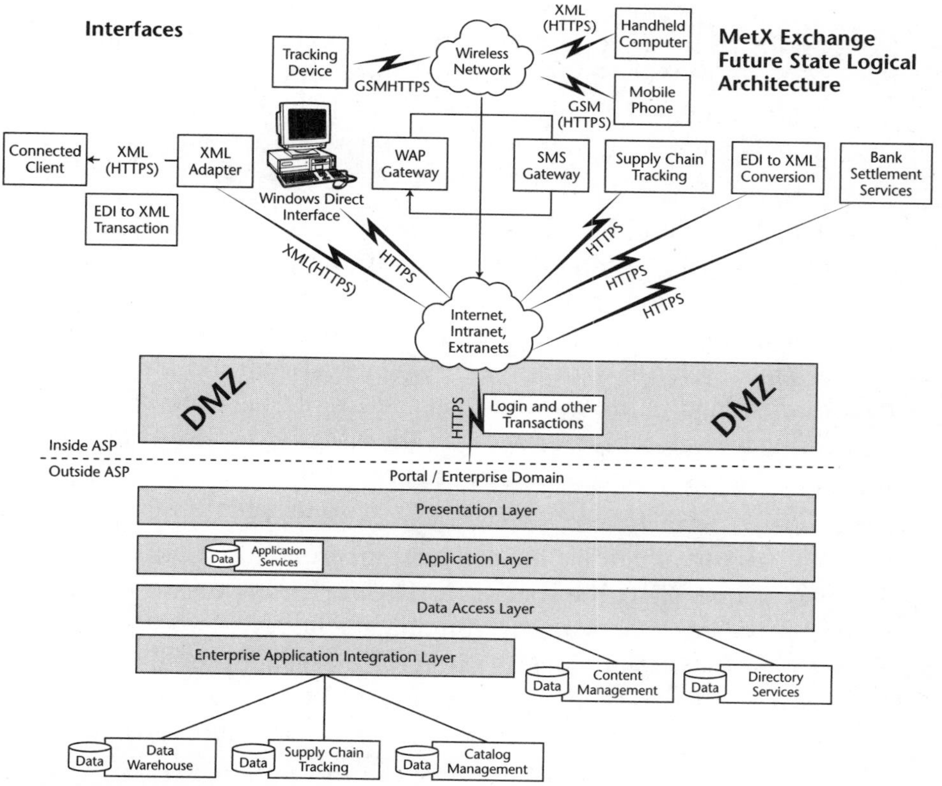

Figure 11.2 Wireless exchange architecture.

Figure 11.2 represents the logical information systems architecture that shows how each of the various layers of the application will communicate with each other and ultimately perform the required functions to support the business requirements. The diagram is *not* intended to be delivered to a development team to begin coding the application. It is, however, intended to provide a baseline framework for the architecting team to understand how information flows from each system layer of the application. Just as a building architect must draw a design of a building's structure, so must a system architect draw an initial design of the application he or she is going to build. Let's review each of the various layers of the design in Figure 11.2 to help you understand what each component is doing.

- **Front-end interfaces.** The front-end interfaces (clients) are the doorways by which users of the system access the main application. In the case of the wireless exchange, the front-end interface(s) can be many things: desktop PCs, mobile devices such as PDAs or WAP phones, and custom devices built specifically for the wireless exchange such as ruggedized handheld computers. The exchange front-end is an Internet portal based thin client application running on any standard Web browser such as Internet Explorer or Netscape Communicator. Each client accesses the exchange network through a secure HTTPS connection through a wireless network that ultimately connects to the client and back-end interfaces.
- **Client and back-end interfaces.** The reference to client and back-end interfaces can be somewhat misleading. When we call a "back-end interface" a type of interface, we are talking about a back-door mechanism for entering the wireless exchange. Traditionally, "back-end" systems refer to data servers, legacy systems, and other applications designed to process the transactions of the "back-office." The two primary back-end interfaces are the XML (Extensible Markup Language) adapter and the connected client. The following is a description of the components that make up these interfaces:
  - **XML adapter.** Handles the conversions from any number of data formats into XML, the specific XML data format used by its back-end processes. For example, it manages the XML processing required to translate EDI function calls. The adapter also handles back-end clients, such as an exchange customer who wants to access back-end data. It performs XML to exchange XML transformations using XSL-T to ensure a uniform data format throughout the system.

- **Connected client.** Any client (a client here is simply the organization, not a system) that has an existing relationship with the wireless exchange to access data from the ERP back-end and to potentially access business information systems such as the ERP Business Warehouse. For instance, if Acme Metals is a member of the exchange community and would like to view accounting/purchase order reports on a regular basis, the company would deploy a connected client that would communicate with the XML adapter.
- **Bank settlement.** The last back-end supply chain and logistics component is the bank settlement interface. As the component's title suggests, someone must be paid when all is said and done. It covers the payment process between carriers and exchange members. The bank settlement tool integrates with the supply chain tracking system to automatically perform financial transactions as orders are placed and filled. Transactions are processed via a flat file.

These interfaces are the primary mechanism for exchange members to access Enterprise Application Integration (EAI) and ERP components, such as the ERP Business Connector.

- **Web components.** The exchange design would not be complete without Web components. One of the most critical layers in the application is the DMZ, or demilitarized zone. The DMZ defines an environment that is more secure than the Internet, and that is located within the domain of an application service provider (ASP) or a corporation. Generally, the DMZ holds servers that provide functionality, but not data. You'll also note that the line separating our front-end interfaces from our Web servers and ultimately the back-end is labeled "inside ASP" and "outside ASP." Obviously, an ASP model was used in this architecture, but that is certainly not a requirement. The enterprise machines in the DMZ are connected to the Internet, so they are more likely to be attacked by hackers. The whole concept of the DMZ is to limit the damage that unwanted access can cause—hackers on those machines are blocked from sensitive data, which is hidden behind enterprise firewalls. A key service that the Web components layer provides is Web connectivity services through a Web server. It also houses the content management services to manage Web content, and the directory services for single sign-on and username security.

- **Logistics/supply chain components.** The exchange would be incomplete without the logistical and supply chain functionality that manages purchase orders and maps customer demand to warehouse supply.
  - **Presentation layer.** The services layer that manages the content and presentation of data from the applications and services is called the presentation layer. This piece is generally concerned with dynamic HTML, Active Server Pages (ASP), and any other elements used for presentation to the client.
  - **Application layer.** This important function manages the application logic that directly communicates with the other layers of the system to process business logic, data access, EAI, and so forth. It also houses the application and business logic of the Web subsystem. In addition, it is the main workflow management, commerce, profiling, and personalization mechanism for users.
  - **Data access layer.** This layer is responsible for retrieving the data using specific data access methods for the mechanism used to store the data. In this case, it is broken into legacy system, local data, and directory access.
  - **Enterprise Application Integration (EAI) layer.** Enterprise Application Integration (EAI) is a key process of most any enterprise application. EAI is defined as the unrestricted sharing of data and business processes throughout the networked applications or data sources within enterprise. It's very rare to design a systemwide architecture without considering how other IT components of the organization will fit into the mix. Most large enterprises have invested a great deal of time and money in existing systems that were originally designed to run on a stand-alone basis. Some of these systems, say legacy accounting or payroll systems, are very large and can contain terabytes of data, whereas others are small, such as a Web server running very specific Java applets that happen to be critical for day-to-day business functions.

Without getting too much more technical, defining the solution for an enterprise application of any significant complexity requires experienced business, technical, data, and application architects. This section only provides a bird's-eye view of the solution to the wireless exchange. For a more detailed explanation of each of the services described, please read *Cap Gemini Ernst & Young Guide To Wireless Enterprise Application Architecture* (Wiley Publishing, Inc. 2001).

Wireless Sales Force Automation

The next case study examines a sales force automation application that was wirelessly enabled to enhance the productivity of our client's field sales force. Before jumping into the subject matter, a few words on customer relationship management and sales force automation are appropriate. In light of an increasingly competitive business environment, many of today's companies have turned to customer relationship management (CRM) tools to more closely bind the customer base to the organization and its offerings. The goals of CRM are to facilitate customer acquisition and increase customer retention, satisfaction, and value. In parallel to companies' realizing the need to optimize interactions with their clientele, customers have become increasingly demanding when it comes to receiving service; they not only expect around the clock access to information that answers their questions, but also expect to be able to interact with a company through whichever channel they may choose, be it via phone, fax, e-mail, the Web, or in person.

CRM tools embody the technology solution that addresses the competitive challenges of a company and the expectations of customers alike. These tools combine three business functions that center on customer interactions: Marketing/Campaign Management, Sales Force Automation, and Call/Contact Center operations. Marketing and Campaign Management operations entail the development and management of direct mail, advertising, telesales, and Web-based selling efforts within an organization. Sales Force Automation is concerned with equipping the field force with the tools it needs to effectively and efficiently maintain and grow the business. Such tools might include applications to identify cross- and up-selling opportunities, or to manage customer files and account histories at the level of the individual sales representative or across the entire organization. Call/Contact Center solutions entail the management of interactions with a customer via multiple channels, especially the phone and the Web.

Remember the customer experience life cycle introduced earlier in this book? At each stage in the process, a customer can interact with the company whose products and services he or she is considering, and vice versa. Figure 11.3 shows the several stages in the life cycle and the features a good CRM solution can provide at each stage.

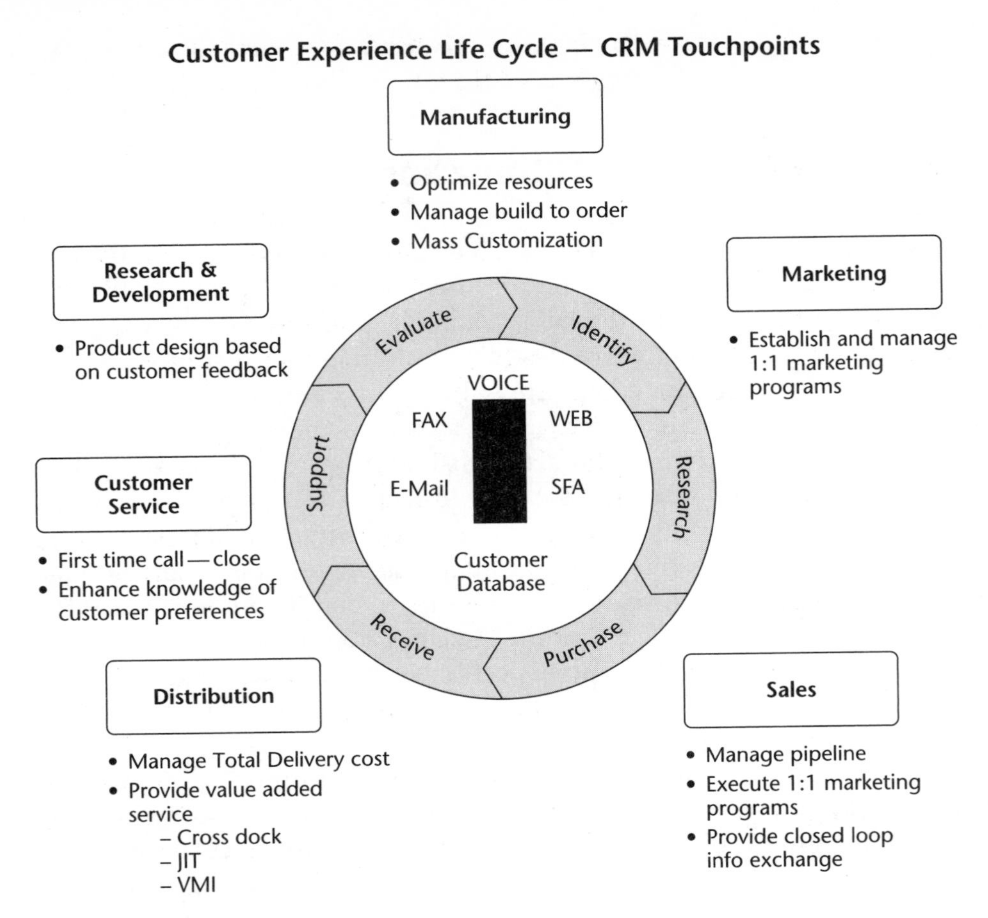

Source: CGE&Y Analysis

Figure 11.3 Customer touchpoints along the experience life cycle.

Conceptual Business Scenario

The case study we will use here represents a wireless application development project we conducted for a client in the radio broadcast industry. The client, a company owning and operating several radio stations around the country, was interested in streamlining business processes associated with selling airtime. Advertising revenue constituted the main source of revenue for this company. Yet, selling airtime across multiple radio stations was a highly complex process. The company owned multiple stations, in several major metropolitan areas around the country, servicing differing demographics, with each station following a certain format (country and western, dance, classic, talk, and so on). The challenge the company's sales reps were facing entailed the coordination of selling literally thousands of highly volatile airtime slots to national advertisers.

National advertisers, large consumer product brands or retailers, for example, would purchase multiple slots of airtime, to play at various times during a day, on multiple stations, in multiple cities, in multiple states to maximize coverage while minimizing cost and effort. The process is inherently complex if one considers that there would be several advertisers vying for a limited number of highly desirable radio spots during the early morning or evening commuting hours of the day. To make matters more difficult, a radio spot is a highly perishable commodity. If the airtime is not sold, it is lost for good and cannot be returned to inventory. In an effort not to let spots go wasted, then, it comes as no surprise that a station would elect to reduce a spot rate progressively as the broadcast date approached in a last effort to sell the perishable item. Communicating available airtime inventory and fluctuating spot prices to an army of sales representatives across the country posed quite a challenge to our client.

So, for radio advertising sales reps to be successful, they must have accurate and timely information about spot availability and price levels. Access to such information was provided by an application that offered visibility into the airtime schedules of each station served. As soon as a spot was sold, it was removed from inventory. For example, if the 30-second slot at 7:35 A.M., on Monday, October 7 on WABC was sold to company X, it was taken off the list of available timeslots a rep could sell to prevent the same slot to be sold by another rep to company Y. In addition to requiring real-time access to the status of airtime inventory and pricing, account executives were working with an inventory that was highly fragmented. National advertisers, usually represented by their advertising agencies, would be interested in buying airtime on multiple stations around the country in a one-stop shopping fashion (cluster buying), adding to the difficulty associated with the process.

In light of these complexities, it comes as no surprise that our client was looking to us for help. CGE&Y had already developed a Web-based sales force automation (SFA) tool for the company that would facilitate and coordinate multiple sales representatives selling thousands of airtime slots at hundreds of stations throughout the nation. This application was used effectively by reps in and out of the office as well as executives on the go, who were equipped with laptop computers that were synchronized with the company's back-end via dial-up connections at the end of the day. Unfortunately, this process did not provide the reps with the real-time information they needed. Although the system did streamline the process, the availability and pricing information on the reps' computers was only as fresh as their last synchronization.

The natural progression was to move the application to a mobile device to improve the timeliness of airtime inventory updates. Figure 11.4 illustrates the solution.

Using the mobile device allowed the inventory to be updated in real-time, ensuring that there would be no conflicts between individual sales reps accidentally selling the same slots to different customers. The mobile application provided the following functionality:

- **Proposal creation.** This allowed a sales representative to prepare a proposal that contained detail surrounding the marketing message, airtime length, pricing, scheduling, locations covered, delivery details, legal terms and conditions, and so forth. Specifically, the proposal creation process entailed a number of screens that allowed the rep to select an account, select the day(s) the message is to be broadcast, specify the U.S. states covered, find all the stations in a given geographic area that satisfy the advertisers requirements, select the target demographics, station format, specify the day part, pricing and other parameters. Finally, the completed proposal was forwarded to individual station managers who had to give approval before the proposal could be finalized and turned into an order.
- **Station filtering.** This allowed the sales reps to quickly focus on the stations that were most desired by the advertiser, filtering the entire inventory of stations by location, format, demographics, and so on.
- **Routing and approval.** Once the proposal was created, the airtime information would be forwarded to each station manager for final approval.
- **Notifications.** Sales reps were able to send and receive real-time alerts to and from station managers and other representatives. In addition to static alerts, the solution allowed for instant messaging sessions, during which reps could engage in conversations with sales directors (to obtain approvals for special pricing, for instance), station mangers, and colleagues.
- **Order management.** This was provided for the management of all orders created by a rep. For example, the solution allowed reps to print the entire order while still at the client site by sending the document to an IP-based printer.

- **Account management.** This offered the ability to maintain account information, including administrative contact information, past order information, customer service issues, payment issues, and other customer intelligence.
- **Reporting.** This provided standard and ad hoc reporting on accounts, proposals, orders and the like.

The wireless solution addressed many of the challenges associated with scheduling complexities and inventory volatility. The sales reps benefited from the application by being able to accelerate the turnaround of a proposal from generation through approval and order finalization due to the faster response times from station managers. This allowed them to service more accounts in the same amount of time, increasing productivity and earnings potential. In addition, the company's sales force was able to realize higher average revenue per account due to the solution's streamlined user interface that allowed sales reps to quote multiple rates across markets more quickly and accurately.

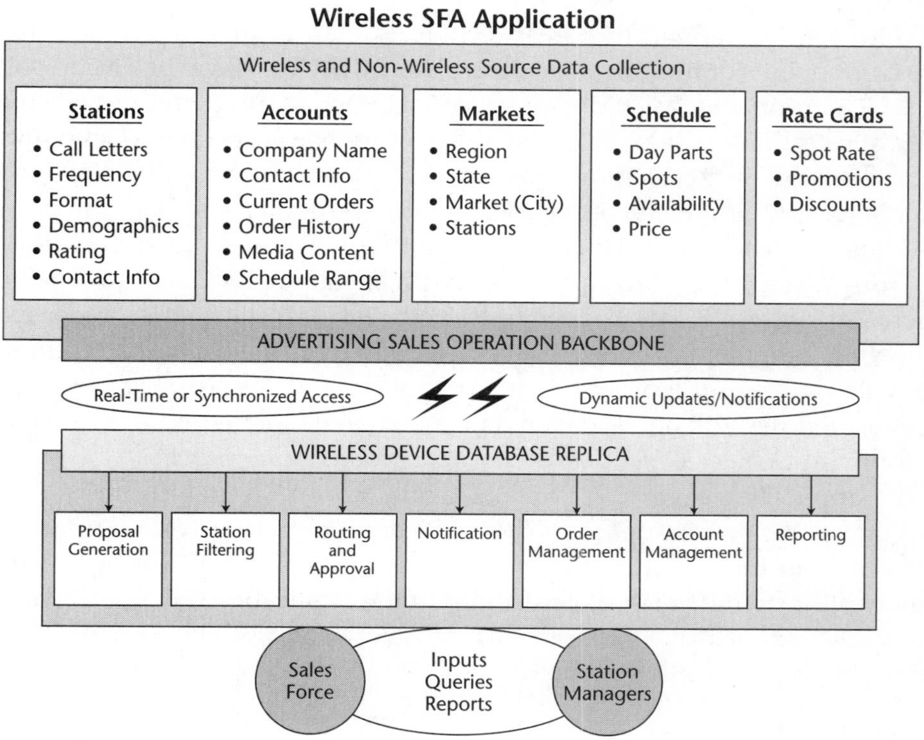

Wireless SFA Application
Wireless and Non-Wireless Source Data Collection
Stations: Call Letters; Frequency; Format; Demographics; Rating; Contact Info
Accounts: Company Name; Contact Info; Current Orders; Order History; Media Content; Schedule Range
Markets: Region; State; Market (City); Stations
Schedule: Day Parts; Spots; Availability; Price
Rate Cards: Spot Rate; Promotions; Discounts
ADVERTISING SALES OPERATION BACKBONE
Real-Time or Synchronized Access
Dynamic Updates/Notifications
WIRELESS DEVICE DATABASE REPLICA
Proposal Generation
Station Filtering
Routing and Approval
Notification
Order Management
Account Management
Reporting
Sales Force
Inputs Queries Reports
Station Managers

Source: CGE&Y

Figure 11.4 Wireless SFA application.

At the same time, our client was able to reduce the number of oversold situations, where the same spot was proposed to multiple clients. Because spot availability was updated faster and communicated to the field near real-time, unsold inventory near its expiration date could be reduced via ad hoc sales campaigns. The electronic nature of the solution resulted in a reduction of order entry and pricing errors, which led to improved station performance.

Last, but not least, our client's customers received added value from being able to access to all of our client's radio stations for cluster buying purposes, and receiving faster confirmations of their orders leading to quicker proposal finalizations.

Defining the Solution

In defining how the wireless components of our SFA application that needed to be built, it was important to understand how the current state (prewireless) of the existing application functioned. The wireless sales force application is a perfect example of how wireless services are an *extension* to the existing technical infrastructure as opposed to a complete rebuild. Without going into great technical detail of this application, the key components of the current state are essentially based off of a Microsoft COM (Component Object Model) based architecture. If you recall from the previous section, each layer of the architecture had a specific function for the complete solution. Those areas included presentation, application, and data layers, all of which communicate with each other and pass information between each other to deliver the services to the client. To continue that discussion, the SFA current state presentation layer is built using Microsoft Active Server Pages (ASP) and DHTML (Dynamic Hypertext Modeling Language). There are several COM+ modules that were required to create orders, fulfill orders, create proposals, and so forth. Each of these COM+ modules will not be covered in any great detail in this book, but can be found in *CGE&Y Guide to Wireless Application Architecture.*

The application layer components, again, are the actual business logic that makes the system operate. The application objects are called upon by Active Server Pages that reside in the presentation layer. These pages ultimately perform the business operations of the application. Finally, the data services layer performs the necessary calls between all the layers to obtain the necessary data called upon.

The future state of the wireless SFA application was designed as extensions to the current state. Again, typically in the wireless world, it's not necessary to redesign an IT infrastructure to support mobility. It is, however, important to understand how each of the new application service communicates with each other to obtain the desired functions. Figure 11.5 illustrates the future state design of the wireless SFA solution.

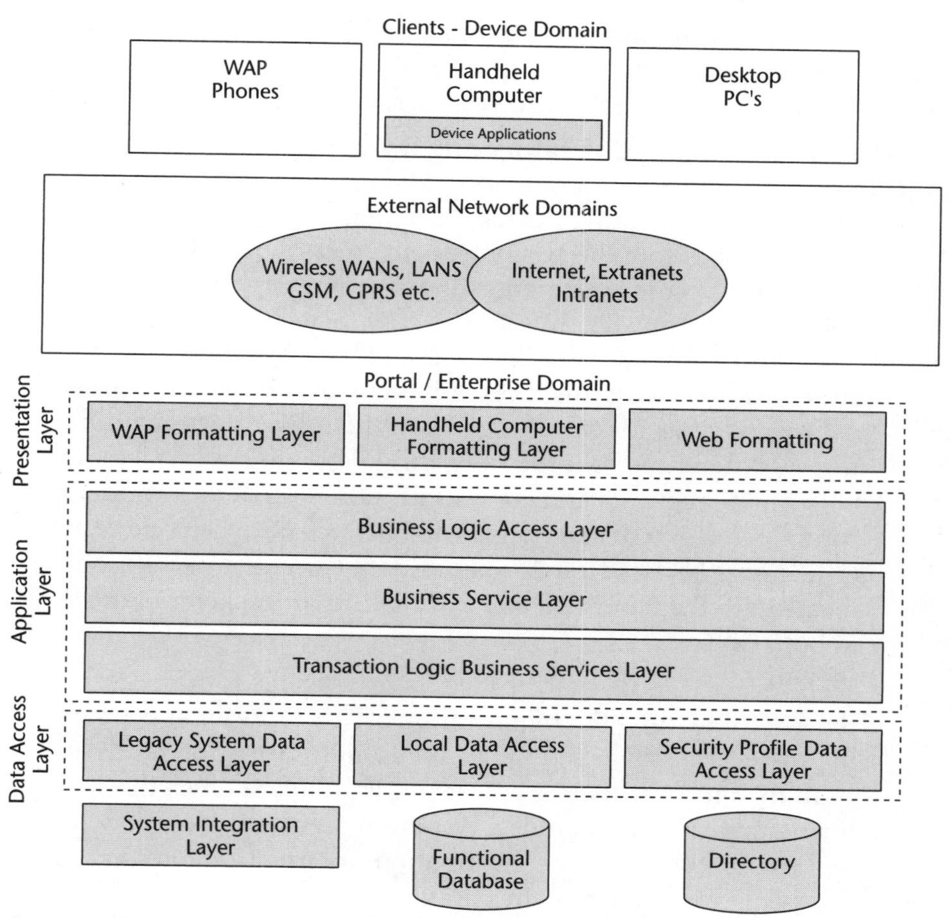

Figure 11.5 Future state design of the wireless SFA application.

The components of the future state can be described as follows:

- The presentation layer has been broken into three pieces:
  - **WAP (Wireless Application Protocol) formatting layer.** This layer is responsible for formatting reading the request from a WAP browser and creating the WML (Wireless Modeling Language) pages required to present information returned from the general presentation layer. This layer knows all specific information about formatting information for a WAP browser and knows how to get and use device specific information (a Nokia phones model X is being used to access the site).
  - **Handheld computing formatting layer.** This layer is responsible for formatting the task of reading the request from a handheld computer and creating the response required to present information returned from the business logic access layer. This layer knows all specific information about formatting information for a handheld PC browser and knows how to get and use device specific information (a Palm Pilot model X is being used to access the site). It is likely that this layer will have to be broken into individual components for each type of handheld PC due to manufacturers' differences. Additionally, this layer will have to handle both online requests and communication with device-resident components. No distinction has to be made between online and offline here because the device-resident components will communicate with the handheld computing formatting layer over HTTP in both cases.
  - **Web formatting layer.** This layer is responsible for formatting the task of reading the request from a Web browser and creating the HTML pages required to present information returned from the business logic access layer. This layer knows the specific formatting information for a Web browser and knows how to get and use browser-specific information (Internet Explorer version X is being used to access the site).
- The application layer has been broken into three pieces:
  - **The business logic access layer.** This layer is responsible for routing the requests from the client side to the appropriate business logic. We don't want the formatting layers to know much about where the business logic resides and have inserted this layer to ensure that this does not have to be the case.

  - **Business services layer.** This layer is responsible for implementing the business logic. It does not know anything about the presentation layer and where data is located.
  - **Transaction logic business services layer.** This layer knows where to get data, but does not know how to get it. This is the responsibility of the data access layer.
- The data access layer has been broken into three pieces:
  - **Legacy access data access layer.** This layer knows how to access any legacy systems. In this case, the only legacy system is the order management system.
  - **Local data access layer.** This layer knows how to access information in the local database. The local database is used primarily for profiling and personalization, but can also be used for staging data. This is usually done to increase performance.
  - **Security profile data access layer.** This layer knows how to access security information in the directory server.

The future state is a more detailed layering of the application components. The rationale behind this layering is to increase the flexibility and scalability of the solution. You can see in the description that each layer has a specific purpose and has minimal information about the other layers. For example, the business services layer has no knowledge of how to access data, or what a WAP phone is. This makes the solution more scalable because the components can be broken down further than in the previous example. It also makes the solution more flexible, mainly because you don't have to change all layers to make modifications to a legacy system or a presentation layer. Going back to our example, we could now implement access via another wireless device without having to change the business services layer.

Mobile Management Information Dashboard

This Mobile Management Information Dashboard case study illustrates how CGE&Y took a commonplace application, a standard management dashboard, and ported it to the mobile environment. As a result, our client's executives and other mobile representatives were not only able to use their devices for personal information management, communications, or to work with office applications, but also to gain insights into key performance indicators while on the go.

Let's start with a brief overview of what a management dashboard does. The management information dashboard provides company executives with critical data and information, or key performance indicators, regarding the organization's operations. Executives who would use such an application range from the most senior C-level officers all the way to leaders of specific business units or functional departments. Any business leader who must maintain oversight of various business processes within his or her organization can benefit from having instant access to key performance indicators that reflect the "health" of the organization to be monitored. Key performance indicators that are commonly included in a dashboard application include categories such as financial, operations, employee productivity, and customer satisfaction indicators, as illustrated in the following list:

- *Financial metrics* can include cash flow, sales or revenue growth, market share, burn rate, or accounts receivable collections.
- *Operations metrics* can involve production processes such as equipment utilization, up- and downtime, scrap, rework, or defects per million units. Other operations-related measures surround warehousing and measure inventory turnover, spoilage, picking accuracy, inventory accuracy, order cycle times, returns, and similar categories. Last, the category also includes distribution, with measurements relating to routing effectiveness/efficiency, scheduling, load optimization, on time deliveries, and the like.
- *Employee productivity metrics* frequently include staffing levels, retention, attrition, sick days, on-time arrivals, time and expense reporting, vacation schedules, and so forth.
- *Customer satisfaction metrics* can include returns, call center contacts, time to resolve, satisfaction ratings, number of product inquiries, account growth, or account attrition.

Usually presented in a highly graphical format, the dashboard functions as a gauge that allows an executive to quickly identify problems and take corrective action. The dashboard avoids deluging the user with data, but instead presents information into critical business processes via select performance metrics such as Defects per Million, Inventory Levels, and Warranty Calls. Note how the color-coding in Figure 11.6 allows the dashboard user to quickly assess the situation.

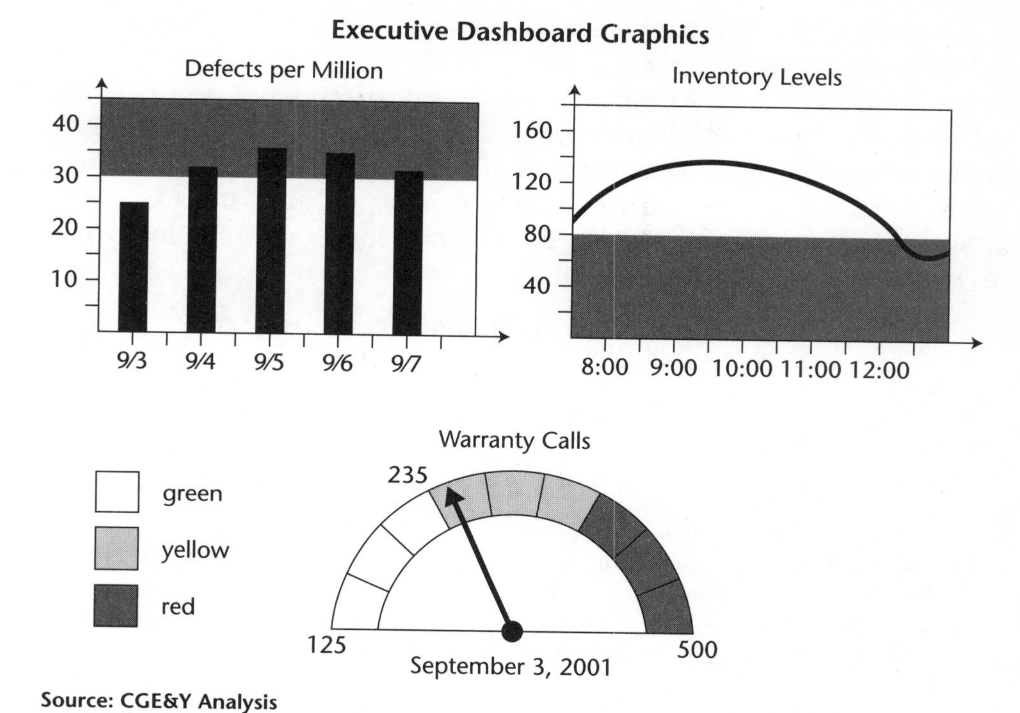

Source: CGE&Y Analysis

Figure 11.6 Executive dashboard graphics.

Conceptual Business Scenario

Our Consumer Products, Retail, and Distribution practice provided the case study that involved the deployment of a mobile management dashboard. The client, a United States–based toy manufacturer with production facilities in Asia, was expanding rapidly throughout the States and Canada. The company's domestic operations involved primarily the design, marketing and distribution of action figures and accessories, whereas production was performed abroad. The finished product was shipped back to the United States, where it was channeled to several warehouses/distribution centers around the country from which retailers were serviced.

Because the organization had grown rapidly over the past few years, the company's executives found themselves increasingly on the road to effectively maintain the company's warehouses/distribution centers and customer base. The growing size of the operation and its geographic dispersion spawned the need for a mobile application that would allow an executive to access a variety of applications and performance metrics while on the go. Our client decided to provide such functionality via a wireless device to its top executives and a certain group of key employees, including traveling sales directors, regional managers, and the sales force. The application provided these highly mobile individuals with the following functionality:

- **Communication.** Voice and data communication with the office in real-time, including e-mail
- **Productivity.** Engaging in personal information management such as calendaring, contact lists, and time and expense reporting, as well as working with office applications such as word-processing, spreadsheet analysis, and presentation programs
- **Information Access.** Being able to tap into legacy systems, the corporate intranet, and external Internet-based information sources for on-demand information, or reviewing corporate key performance indicators pushed to the device on an exception basis

Figure 11.7 illustrates the application. Note the various back-end systems that act as the sources for data and information to be displayed on the mobile device.

In general, this simple yet very effective application allowed its users to make better decisions, faster. The wireless solution alerted our client's designated dashboard users on an exception basis. Whenever a certain metric, specified during the development of the solution by the various user groups involved in its design, was deviating from the standard, a system-generated message would be sent to the mobile device. Such deviations could be positive or negative in nature, exemplified by a large sales contract that required immediate special pricing approvals, or the failure of production equipment that required immediate attention regarding its repair.

In addition to system-generated alerts, the solution allowed for instant communication with dashboard users on the go. Although this functionality is nothing special in our age of cellular communications, it acquires new meaning when such communication can be combined with the aforementioned actionable alerts. The operative word here is "actionable." Being able to receive an alert, say, about an equipment malfunction, and then being able to work with that alert by responding to it, forwarding it, or invoking a set of actions significantly enhances the value of the alert. For example, the alert may entail a message that a certain piece of equipment is malfunctioning. A static alert simply relays this message, whereas an actionable alert offers additional workflow steps. Specifically, an actionable alert may include the option to notify a group of individuals that are associated with the machine (operators, supervisors, maintenance crews), or it may allow the receiver of the alert to shut down the equipment altogether using the device the alert was issued to.

Last, the solution allows the user to stay productive while on the go, and offers access to intranet and Internet-based information sources. Because of this, the solution acts like a mobile office that has been enhanced with functionality that provides real-time visibility into the organization's operational metrics. A powerful tool indeed.

Executive Dashboard Application

Source Data

ERP Stations
- Financials
- Customer Profiles
- Employee Data
- Sales Data
- Procurement Data
- Production Data
- Distribution Data

Administrative Support
- General Support
- Custom Reports
- T&E Reporting

SBU Management
- SBU-Specific Key Performance Indicators

Intranet/Internet
- Corporate Knowledge Management System
- Industry News, Trends, Analyses
- Customer and Competitor Intelligence

ENTERPRISE BACKBONE

Real-Time or Synchronized Access

Dynamic Updates/Notifications

WIRELESS DEVICE

Communication: Voice, Data

Productivity: Personal Information Management, Office Applications

Information Access: On-demand Information, Exception Alerts

Executive

Source: CGE&Y Analysis

Figure 11.7 Executive dashboard application.

Defining the Solution

The mobile office solution architecture is essentially broken into four buckets of application functionality:

- **Office automation.** This area generally includes your Microsoft type of applications such as Word, Excel, PowerPoint, Access, and so on, which covers word-processing, spreadsheets, graphical editors, and others.
- **Communication**. This category covers typical heavily used applications such as e-mail, chat, videoconferencing, and online meetings.
- **Information management.** Personal information management (PIM) applications have long been a foundation for the office environment. Functions including calendar management, document management, Internet/intranet/extranet access, and knowledge management are all staple applications in this environment.
- **Intercompany data.** A very broad category that constitutes accessibility of company back-end data, such as data warehouses, ERP reports, financial data, customer data, employee data, and so forth.

Mobile office functionality is constantly expanding to include applications that were typically wired to the desktop, but as wireless networks bandwidths are exceedingly becoming faster so are the accessibility features.

Documenting the complete infrastructure of the mobile office executive dashboard is a large endeavor and will not be necessary to help you understand the solution definition. So in its place let's take a look at the future state analysis diagram (see Figure 11.8) and review the primary components from a high level.

The main components or blocks can be defined as follows:

- **Common office platform.** These applications generally consist of your standard office applications and traditional front-end tools such as Web browsers, word processors, utilities, graphics viewers, and even ERP-facing tools. The common office also includes advanced office services such as personal databases, graphical editors, and possibly custom internal applications for time and expense tracking, personal information managers, groupware, and others. Finally, the critical piece that integrates the standard and advanced clients is some sort of local area network, or local office equipment. This component gives our mobile office users a connection to the back-end services that are so critical to the collaborative design of the mobile office.

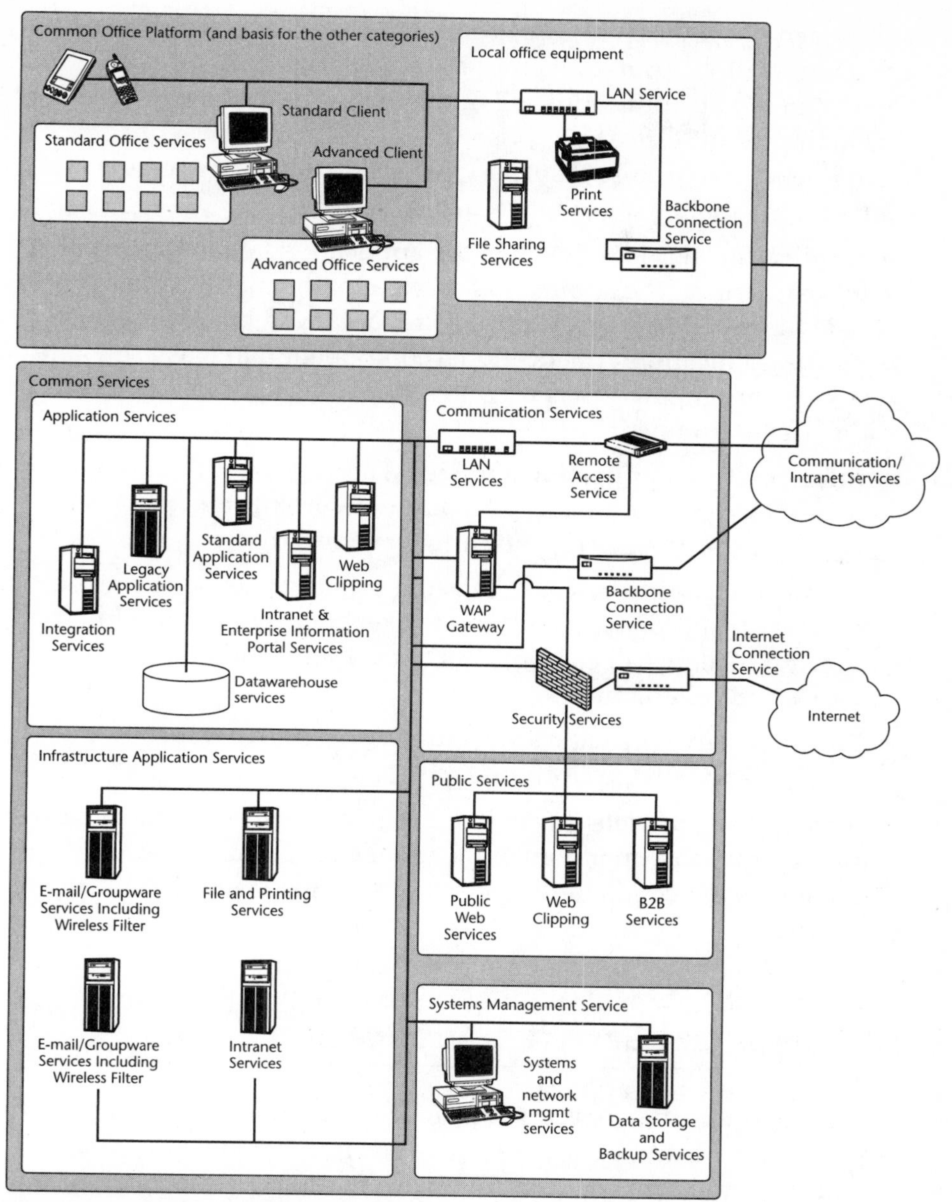

Figure 11.8 Mobile office future state.

- **Common services.** The true backbone of the mobile office infrastructure resides in the common services components. Each of the services within the common services component plays an important role in the mobile office architecture. The reason that all of these components together make up common services is that they are common to all the users. Whether the application calls for a ERP tool, or an e-mail application, all of the tools are common to everyone in the mobile office enterprise. Within the common services component, several subsets of services reside in the application that allow for the mobile office functionality, those are:

- **Application services.** Just like the application layer of all *n*-tiered infrastructures, the application services of the mobile office provide the EAI, ERP, legacy, portal, and data warehousing services for the application. An additional component that was used here is Web clipping, which is used to serve Web pages in a mobile device environment.
- **Infrastructure application services.** These services are required for e-mail/groupware functionality, network file and print services, directory services, and intranet connectivity.
- **Communication services.** Act as the primary backbone connections for LAN connections, remote user connections, and Internet firewalls. Allows all components of the mobile office to communicate and collaborate as one.
- **Public services.** Provide access to external services such as the Internet and B2B applications.

- **Systems management service.** Used primarily by the systems administrators and operators of the mobile office infrastructure. Components include data recovery and systems management console services.

Mobile Financial Services Application

The financial services industry was one of the first to move operations to an online medium. Traditionally set up as electronic networks, securities trading, for example, has made a relatively fast progression out of the offices of full-service brokerages and onto the Internet to be accessible to millions of consumers. In addition to trading, we currently witness the sector adding banking and insurance services to combine with trading into a full-fledged offering of financial services.

The convergence that results in the three types of services being bundled is due to deregulation, the consolidation of the companies providing these services, and global expansion. Whereas European financial institutions have been able to offer a combination of financial services for a relatively long time, the trend is relatively new in the United States where the Gramm-Leach-Bliley Act (GLBA) of 1999 opened the door for consolidating such services to be offered by so-called financial holding companies.

In addition to converging services resulting from deregulation, the U.S. market is characterized by providers merging and aligning with each other in an effort to form powerhouses that can withstand the increasing domestic and foreign competition. The name of the game is customer retention; if financial holding companies can manage to bind customers to their offerings from their earliest contacts via banking services to trading and obtaining insurance coverage, these institutions hope to effectively lock in their clientele. By taking care of all aspects of a consumer's financial needs these organizations calculate they will be able to increase revenues and customer loyalty, having raised switching barriers to levels that would make it undesirable for consumers to obtain financial services and manage their accounts by having to work with multiple, nonintegrated providers.

Last, the GLBA, which allowed for the bundling of financial services, also opened the U.S. market to foreign companies in the industry. These international financial powerhouses view the American market as a huge opportunity to expand their reach. At the same time, the world has become the new playground upon which American companies may elect to step. But to do so, they must quickly obtain critical mass; an effort illustrated by the increasing merger activity and alliances within the industry.

In light of these significant trends in the financial services industry, you can understand how these companies are constantly evaluating new technologies in their quest to improve customer service. First, there was the Internet, now there is mobility—the industry's new frontier.

The Situation

The case study we are presenting here comes from one of Microsoft's clients who was willing to selectively invest in new technologies in an effort to learn how they could add value to the organization. Scotiabank is one of North America's premier financial institutions and Canada's most international bank. With 49,000 employees, Scotiabank Group and its affiliates serve about 10 million customers in some 50 countries around the world.

The automotive flooring business is one of underwriting the inventory of new and used vehicles for automotive dealerships, requiring the bank to perform periodic audits to ensure that the vehicles on the dealer lot correspond to those listed in bank records. Standard practice was to use paper listings of the dealer inventory. The auditor walked the lot and checked off vehicles, using the 17-digit Vehicle Identification Number (VIN) as identification. The auditor then met with a dealership representative to reconcile discrepancies. Last, the auditor sent the audit results by postal mail or fax to its Scotiabank headquarters, where results were hand-keyed into an electronic flooring system database.

This audit process had several problems, including:

- Error rates would go as high as 20 percent.
- The process was time-consuming.
- Delays in obtaining paperwork created problems. Once an auditor requested a dealer inventory list, there could be a delay of as much as several days before the inventory list was received and the audit took place. During that time, the dealer typically sold several of the vehicles on the list and acquired new ones. Because the auditor was working with a now-outdated list, the reconciliation process took extra work.
- Auditors did all work by hand and on paper. There was no automated method to capture audit results for tracking purposes. The results were compiled using paper reports and were hand-keyed into a database.

Auditors needed a way to obtain completely up-to-date inventory lists, as well as to more efficiently and accurately collect and report findings. The bank wanted to be able to give its auditors the tools they needed—on a handheld device that offered wireless capability. But because many dealerships were in rural areas without wireless service, the application had to perform well in a disconnected mode.

The Solution

Scotiabank, working with Microsoft Consulting Services (MCS) Canada, created a mobile data capture and storage application that is based on the Microsoft Windows–powered Pocket PC Phone Edition. Developing a prototype took two full-time and two part-time developers only 20 days. The application has been tested and, as of the writing of this book, will soon be rolled out to all of Scotiabank's flooring auditors across Canada.

The new application improves the auditing process at every stage:

- **Stage 1: Before the audit.** To obtain a dealer inventory list, the auditor simply makes the request on the handheld device and synchronizes the information, by using either the wireless capability or the device's cradle. The device connects to a Web service and downloads the inventory list.
- **Stage 2: During the audit.** The auditor can input VINs either by using the on-screen keypad or by speaking numbers into the device. The device then gives an audible response. Speech-recognition software from a third-party vendor makes this possible.
- **Stage 3: During audit reconciliation.** The application automates certain aspects of the reconciliation process, using drop-down lists to simplify data entry and, when necessary, prompting the auditor to obtain additional information and documentation. For example, when an automobile has been sold, the device prompts the auditor to view the contract and record the license plate number.
- **Stage 4: After the audit.** The auditor synchronizes the device, and audit results are transferred to the bank's Automobile Leasing and Flooring (ALF) system by way of a Web service (see Figure 11.9).

Scotiabank already had developed its ALF system, basing it on the Microsoft .NET Framework, to give Web access to dealer inventory lists, so using the .NET Compact Framework to develop a handheld application that consumed this information was a natural next step.

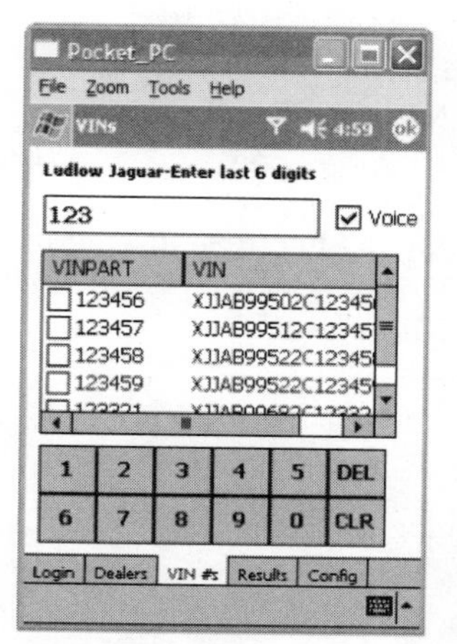

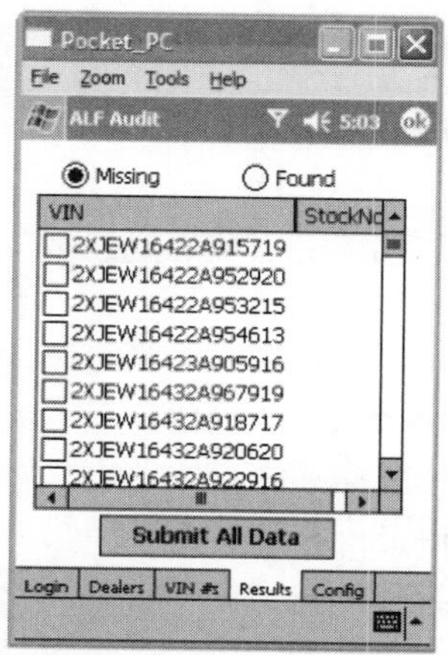

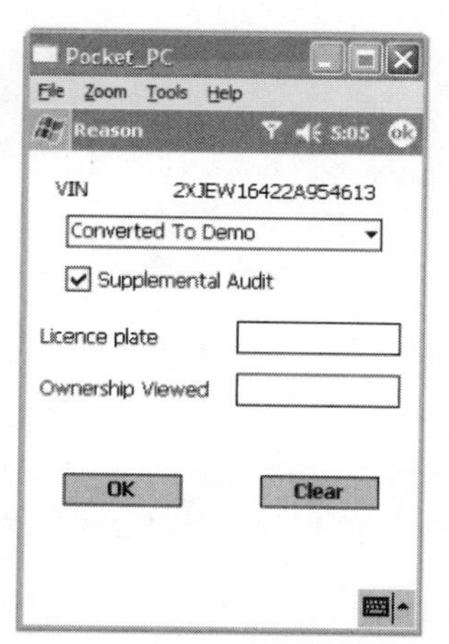

Figure 11.9 The Scotiabank handheld auditing application.

The ALF system was built on the .NET Framework using C# and uses IBM Netfinity servers with SQL Server 7.0 as the database. The server operating system is Windows 2000 Advanced Server with Internet Information Services 5.0. ALF relies on a back-end IBM mainframe, which daily stores the data and processes payments from, and updates inventory data to, the ALF system in batches. Figure 11.10 shows the application's high-level architecture.

The Benefits

The handheld auditing application possesses many advantages unparalleled by other applications. It improved the auditing process at every stage, simplifying data capture and reporting, saving time, and improving customer service. Most importantly, it works all the time, even in disconnected mode. Scotiabank required an application that would take advantage of a wireless connection if it were available, but not be dependent on it. The richness of the Pocket PC environment allowed the team to write an intelligent application that works with or without a wireless connection.

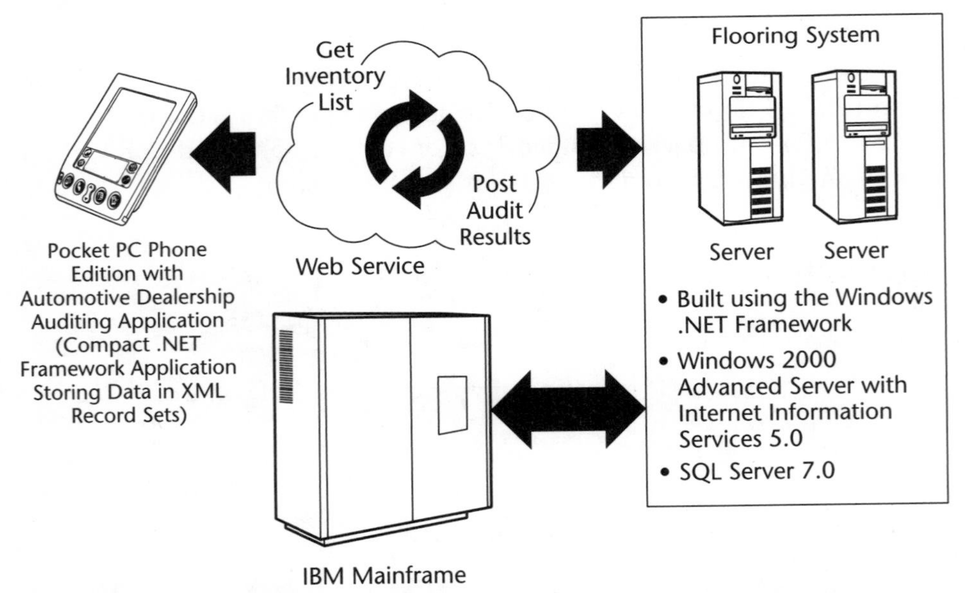

Figure 11.10 The Scotiabank flooring system dealership auditing application.

Streamlined Data Capture and Reporting

Because the Pocket PC provides the ideal audio input medium, developers decided to use third-party speech recognition software. The bank had excellent accuracy with speech recognition, which gives auditors a very functional and useful alternative to tapping numbers on a keypad. The handheld application also greatly improves the process of reconciling audit discrepancies and reporting findings. The handheld application simplifies and standardizes the audit process by providing drop-down lists and prompting the auditor to record specific data about a particular vehicle. Best of all, it provides—in place of a stack of forms—a single place to record all information. The auditor posts audit results directly to the ALF system database with a single synchronization.

More Meaningful Business Intelligence

Scotiabank expects to take the next step and transform audit data into meaningful business intelligence, thereby essentially turning data into an asset. The mobile application provides a way to do meaningful trend analysis, statistics, and other reporting that will help managers, auditors, and sales people sell services and make more informed decisions.

Improved Customer Service

Previously, a delay of several days could occur before a requested inventory list would arrive. The new application lets an auditor download a completely up-to-date inventory list to the Pocket PC device immediately before starting an audit. Working with fresh data, an auditor should encounter fewer discrepancies and, therefore, have to do less reconciliation. Scotiabank's auditors typically interact with some of the most time-constrained people at the dealership—the controllers, who are busy processing dozens of deals a day. So the fewer vehicles they must reconcile, the more time they can save the customer, and ultimately provide better customer service.

Reduced Frequency of Errors

With the paper-based system, hand-keying audit reports into the database resulted in an error rate as high as 20 percent. Because the handheld application uploads the audit findings directly to the ALF system database, no hand-keying is necessary.

Robust, Easy-to-Use Development Platform

One of the many challenges companies face in developing applications for mobile devices is learning new tools and development languages that are often required for new environments. The development environment and the transferability of developer skills are very important when deciding a mobile development platform. In the case of Scotiabank, the Visual Studio development tools were used that allow developers to create applications with the same development languages as a normal desktop or Windows developer. This allowed the bank to leverage the same technical and developer skills without a great deal of time spent learning new application developer kits.

The Far-Reaching Possibilities

Looking to the future, the bank envisions banking directed toward a model where bankers go to customers, using mobile computing applications on Pocket PCs or Tablet PCs. This type of mobile banking would require the ability to access central systems without being tethered to them. For example: Standing in the model home at a new housing development, the bank could preapprove someone for a mortgage from a *virtual branch*, on the spot. Having given the buyers the happy news that they qualify, they could begin the mortgage application and also have access to a broad range of other banking services. The software and technology that was used to create this auditing application shows that this kind of service is a real possibility.

Summary

This chapter tried to touch upon some of the very exciting case studies and opportunities that we've seen in our experience in the mobility world. Undoubtedly, you'll find that wireless technology is a very dynamic environment and applications are constantly evolving to keep up with the demands of the mobile consumer. Hopefully, this chapter offers a glimpse of how mobility has changed the way our clients think about new opportunities to service their customer base. In the next chapter, we'll introduce a structured, strategic approach that a company might follow to develop and deploy a mobile solution.

PART Three

Strategy Formulation and Implementation Methodology

CHAPTER

12

Developing a Mobile/Wireless Solution

Jorn Teutloff

Introduction

In the last section of this book, we lay out a high-level roadmap for readers who would like to deploy mobile and/or wireless technologies in their organizations and are looking for a proven approach that leads them along the way. In Part I, we introduced you to mobile and wireless definitions and terminology, presented the drivers that sparked the industry's fire and will fuel its growth, and showcased some of today's mobility hardware and the networks this exciting technology runs on. Part II then introduced the Value Web concept, and explained in a step-by-step manner how to build your own Value Web as part of your mobile technology strategy formulation. The section closed with a discussion of a Value Web CGE&Y prepared for today's wireless industry that introduced some of the companies currently playing in the emerging field. The next chapters jumped into the various types of mobile applications and gave you some very specific examples for applications currently on the market. The section also included several case studies that incorporated real-life projects that CGE&Y conducted for the consultancy's clients. In the last two chapters of the book (Chapters 12 and 13), we will leave you with a high-level roadmap that builds on all the approaches and examples we've looked at up to this point. The goal of these last two chapters is to provide you with the tools you need to embark on your own journey into the mobile technologies realm.

Although it cannot be the aim of this book to make you an expert on mobile/wireless technologies and their implementation in a corporate setting overnight, this chapter provides you with the knowledge about the industry's current state and some specific techniques to apply in your organization to get the mobility ball rolling. After reading Chapters 12 and 13, you will have an understanding of the critical phases of developing a mobile/wireless solution for your enterprise.

Following a Structured Approach

The primary benefits associated with deploying mobile/wireless technologies in a corporate setting include the potential to grow revenues, reduce operating expenses, streamline business processes, enhance the organization's competitive position, and/or improve relationships with the company's stakeholders, including suppliers, alliance partners, employees, customers, and shareholders.

To attain these benefits, mobile solutions must be customized according to a company's strategic direction and unique business requirements. Every organization runs differently and operates within a unique set of environmental parameters that affect which mobile technologies are most appropriate and how to best implement them. Thus, before you can effectively deploy any initiatives, an organization must be clear about where it is headed in the long run and how mobility can best assist it in getting there. It is not uncommon that a company faces a slew of mobile and/or wireless point solutions—all of which provide some benefit. Yet in the light of scarce resources and difficult economic times, organizations must decide which of these potential solutions make the most sense. Arriving at that decision requires a structured approach that takes into account various internal and external realities.

Opponents of this "strategy-by-design" approach favor the rapid deployment of point solutions that address a current issue at hand. Although this philosophy can be very effective at stopping the bleeding—a valid justification for any "shoot-from-the-hip" solution—don't accept this approach as a replacement for a thorough analysis. Looking back just a few years, we all are aware of a multitude of companies that jumped onto the Internet bandwagon without much planning only to fail after realizing that the solutions they built lacked the financial or strategic justifications that a solid analysis and the preparation of business cases could have provided. If there is one thing that we have learned from these companies, it is that unless there exists a clear value proposition, a value proposition that is founded on a solid, fact-based analysis, the risk of failure is tremendous and overshadows the potential for discovering a diamond in the rough.

To illustrate this point, we see a lot of retailers that currently consider making their Internet sites accessible to wireless devices. The reasoning, so it goes, is that customers would value the wireless channel as yet another avenue to do business with the company. Yet, we have strong doubts about this one-size-fits-all approach. For most companies, simply fronting an existing Web site with a wireless interface and making a consumer product available for sale via a cell phone or a PDA does not reflect the unique value propositions mobile technologies offer. What makes wireless unique is its ability to assist mobile users with making time-critical decisions, allowing users on the go to conduct transactions that are simple, and providing users with location-aware functionality. To use an extreme example, why would you purchase a bed frame, laptop computer, or a pair of pants while on the go, say while driving to work in the morning? You wouldn't! But you would be likely to use your mobile device to obtain location-specific directions to find your way to a client's office you haven't visited before, or to react to an actionable alert that notifies you of your afternoon flight being canceled and asking you whether to rebook or cancel the trip.

The point is that without a structured approach to planning, you might jump onto opportunities that later turn out to be inferior to others, or at worst dead ends. Will a structured planning approach completely eliminate going down the wrong path? Probably not. But it will allow companies to make better decisions and discover opportunities that lie beyond the obvious. A solid strategy built upon clearly defined business objectives provides overall direction and guiding principles to select the most appropriate solutions, and forms the foundation for a solid systems design and implementation plan.

The major phases to be followed during a mobile solution deployment include the development of your strategy, the implementation of this strategy, and the ongoing monitoring of said strategy in light of ever-evolving technologies, business requirements, and market realities.

Phase I: Mobile/Wireless Strategy Formulation

The purpose of this phase is to develop, refine, and document a company's strategy regarding the use of mobile and wireless technologies. Specific stages within the strategy development phase include the definition of a mobile/wireless vision, setting the direction, assembling a portfolio of tactical initiatives, developing a high-level deployment roadmap, and creating a proof-of-concept. Significant milestones throughout the process include (A) the initial funding decision that allocates resources to the strategy formulation project, (B) an initiative prioritization exercise after the portfolio of potential initiatives has been created, and (C) the go/no-go decision upon completion of the pilot which could also be viewed as the second round of funding as it clears the resources required for the build-out of the actual solution. Figure 12.1 illustrates the stages within the strategy-formulation phase.

Stage 1: Developing a Vision

At the beginning of every journey there usually lies a vision. You need to formulate, own, and communicate this vision to serve as a beacon that shows the way. Figure 12.2 illustrates the activities involved in formulating a vision.

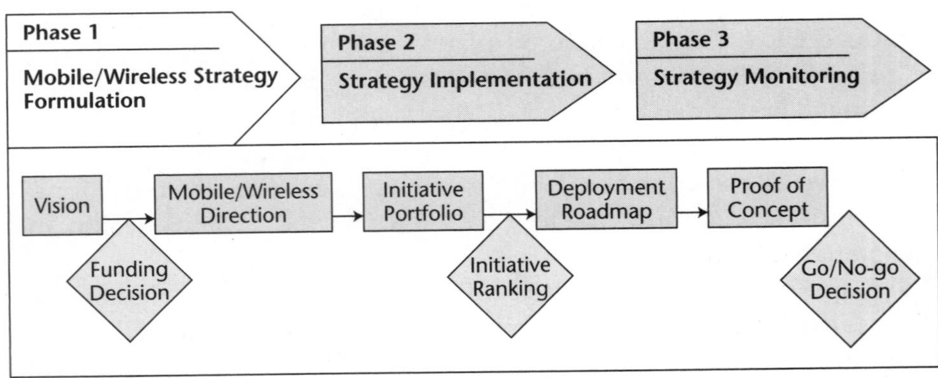

Figure 12.1 Mobile technology strategy-formulation process.

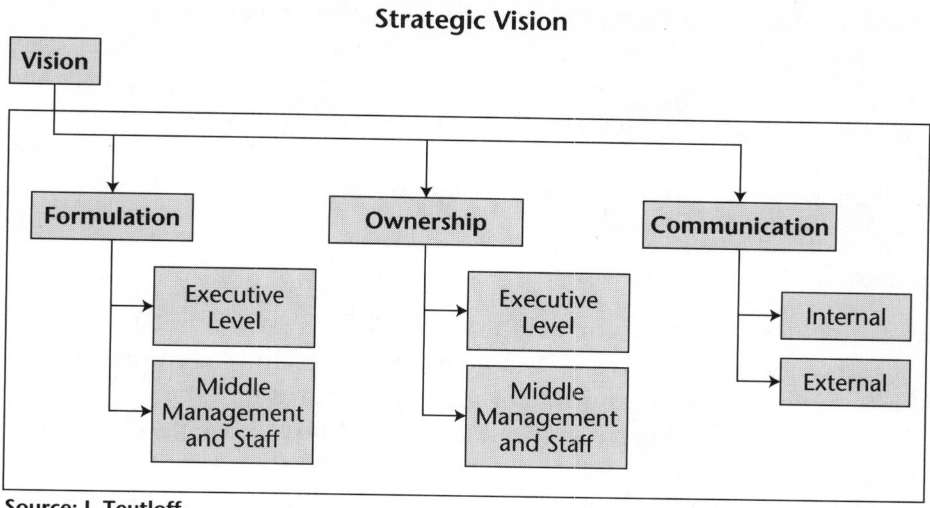

Source: J. Teutloff

Figure 12.2 A strategic vision takes some planning and effort at all levels.

Let's look at each of these activities in more detail.

Formulation

Commonly developed by the organization's senior executives, a strategic vision about mobile technologies reflects cutting-edge thinking about how the deployment of such technologies might assist a company in attaining financial or strategic benefits. If not developed by senior management, it is not uncommon for such a vision to bubble up from the ranks, originating in the middle management layers of a company or formulated by employees involved with the daily operations of the enterprise.

Regardless of where it originates, an organization's vision as relating to mobile functionality embodies a long-term mental picture about where the company will be in a few years and how mobility has affected the company along the way. Such a strategic vision might position mobile technologies as an enabler that will allow for significant enhancements in how the enterprise operates, it might stipulate how these technologies can be deployed in an effort to upset current industry structures and change the game altogether, or it could entail the creation of a an entirely new business.

Regardless of the strategic direction, the mental picture usually reflects a positive attitude about the company in the future and thus serves as a rallying cry behind which the organization's stakeholders can gather to move forward in a united fashion. Although lacking clear details about the specific technologies employed, the vision focuses on features and the user groups that would benefit from implementation. Simplicity and clarity are two characteristics that describe an effective vision; everybody understands the general idea, and everybody can communicate it to others in the organization—usually with just a few sentences.

Ownership

Now, to take this vision as a foundation upon which to build a mobile technology strategy and from which to spawn specific technology initiatives, it has to be owned by the individual or team that makes long-term decisions for the organization. Unless senior management buys into the vision, any ideas about how mobility will serve the organization are wishful thinking with minimal chances of evolving into reality. Let's face it, only when a company's leadership signs up on the dotted line will mobile initiatives have a chance of being implemented; after all, technology initiatives do require the investment of scarce resources. That is not to say that a lack of vision at the executive level means the death knell to all mobility efforts; we have heard about quite a few savvy corporate "intrapreneurs" at the front line level who identify a pressing need and introduce a quick-and-dirty, simple yet effective wireless solution, built in their garages, in their spare time. If these solutions are truly relevant to the organization and effectively address pressing business issues—versus being developed because they provide "cool" features—they can quickly become indispensable, prompting senior management to officially recognize the technology. The point though, remains; only if your organization's leadership is on board will the resources become available for the official development and deployment of a comprehensive mobile technologies solution.

Having said that, it is equally important to gain the buy-in and sense of ownership from those who will actually have to use the technology. The landscape is littered with failed IT applications that were conceived for all the right reasons but failed to create enthusiasm and adoption by those they were built for. Nobody likes for technology to be shoved down his or her throat; so make sure you instill a sense of ownership in the involved user communities very early in the process to obtain their support all the way from strategy formulation to solution deployment.

Communication

One additional factor to consider when talking about vision is how to communicate the big picture, and to whom. As stated earlier, the vision serves as a central theme that unites people in their quest for achieving what the vision sets forth. To ensure that everybody in the organization is on the same page, however, the vision must be communicated effectively. Effectively infusing an organization with the spirit captured in your vision is easier said than done unless you find a way to make your colleagues see why the idea is important for the enterprise and what's in it for them.

Now think about your organization. If you are a member of the senior executive team, has your team formulated a vision for your company? Where do you personally see the company within a couple of years? Does the team's vision, or your own, entail mobile functionality closely engrained in the organization's daily operations? How will wireless allow you to compete more effectively? Are there opportunities to launch a new business that is founded on your original offering but takes it to a completely new level because of mobility? Have you communicated this vision through the ranks in an effort to rally the troops? How effective do you think your communication has been? In other words, if you were to walk the shop floor right now and ask one of your warehousemen about your company's vision, would you receive a concise statement that captures where the senior management team sees the company in a few years from now, or would you receive a blank stare?

If you represent middle management or work in a specific functional area, are you aware of your company's vision regarding mobile solutions? Does your company have one, but it's not communicated effectively? Do you see opportunities, due to your daily exposure to tasks, that would greatly benefit from a mobile application? Have you talked with your peers to confirm this opportunity? Have you thought about a high-level business case behind your vision? Would you be able to summarize in a few bullet points the costs and benefits associates with your idea? Have you written up your thoughts and communicated them to your boss or others in your organization who would be able to take your vision to the next level such as launching a formal investigation of its merits? If you have not, why not?

MILESTONE A: THE INITIAL FUNDING DECISION

Before the first milestone is reached, someone within the organization must assume the role of a project champion to help expose the opportunity to others and to coordinate activities after funding is approved. The project champion is a strong believer in the cause and has made it his or her goal to drive it forward. The project champion can be an individual who was hired specifically for this role, but more often we see current employees taking on this function because they realize the technology's potential for the firm or out of sheer personal enthusiasm about mobile/wireless technologies and their applications. The champion is determined to see the project come to fruition and is willing to go the extra mile to make it happen. At CGE&Y, we have seen such project champions come from any area within an organization, yet frequently they are members of the company's senior executive team or belong to the company's IT department.

The vision is clear, senior management shares it, employees understand it, and the board has authorized the budget to further investigate opportunities associated with mobile technologies. The company has reached the first milestone—the initial funding decision. Having the board and/or senior management sign off on the project makes it official, and provides for the resources—both financial and human—to be dedicated to the project.

With executive sponsorship and funding secured, the project champion is ready to assemble a team to conduct the in-depth analysis presented next. If the team is recruited from within the company, look for individuals well versed in strategic research and analytical decision making. We found that the mobile technology strategy formulation phase benefits most from a team that understands business processes and requirements, is aware of industry trends and leading practices, and is adept at researching and assessing mobile/wireless applications, whether they are deployed within the company's industry or outside of it. If the team is contracted from the outside, make sure that your consulting partner of choice has a deep experience in similar strategic projects and maintains a roster of trained technology specialists to assist with the actual technical deployment of the solution and its integration with your existing systems.

Stage 2: Setting the Direction

Whereas the vision represents the high-level mental image of what mobile technology can do for a company, the next activity sets the direction for specific initiatives. A mobile/wireless direction outlines where the organization is headed and what role mobile technologies will play. Again, this technology can be deployed to improve a company's operations, it can be used to upset the status quo in your current Value Web, or it can be leveraged to create entirely new entities.

When this chapter discusses strategic direction from this point forward, it mainly refers to competitive and functional strategy. Although some readers might use this book to investigate opportunities for setting corporate strategy (the definition of the markets in which their company should be competing), most of our readers will be concerned with using mobile technologies to establish competitive advantage or to streamline internal operations within their organizations. Both objectives can be attained through the formulation of specific, tactical initiatives. Before you develop such initiatives, you need to understand what types of mobile/wireless functionality exists in the marketplace, how they might address your company's business requirements, and which internal capabilities you possess to actually deploy such initiatives.

To develop this understanding, companies should follow a set of activities that simultaneously assess their internal and external environments and develop a Value Web that illustrates the power relationships within the firm's market space. Figure 12.3 lays out the approach.

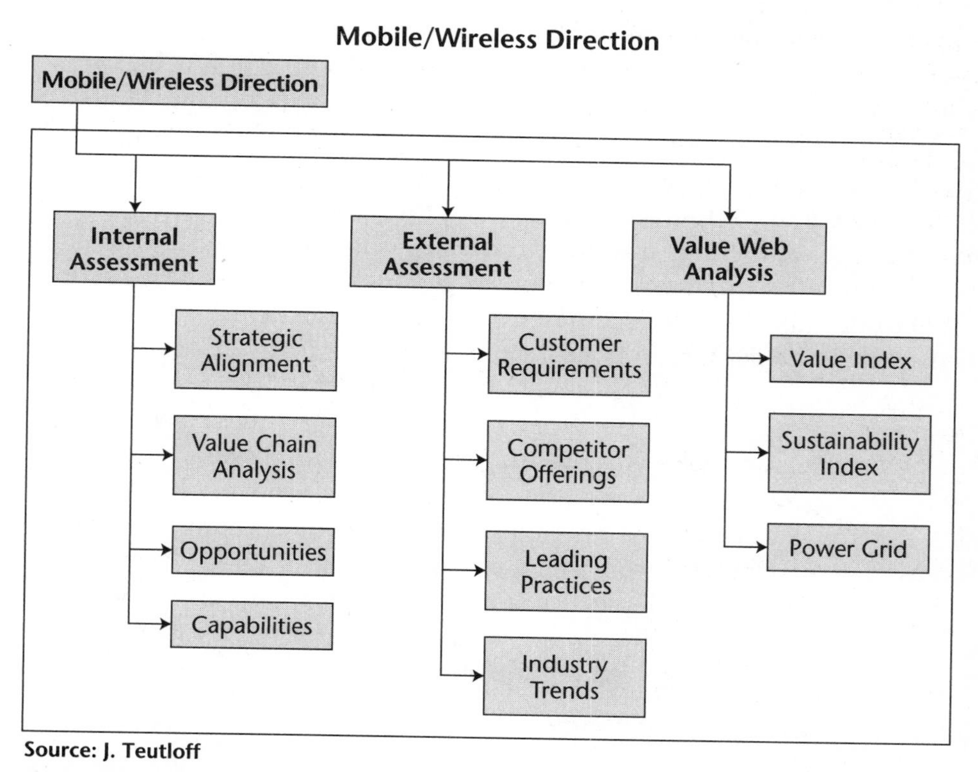

Source: J. Teutloff

Figure 12.3 Mobile/wireless direction.

Internal Assessment

The first activity within the stage of setting your mobile/wireless direction entails the assessment of the organization's internal environment. Specifically, you need to understand what the opportunities are within the organization and what internal capabilities exist to address them. An analysis of your organization will provide you with this understanding.

First, it's probably a good idea to revisit the company's documented strategy; at each level—corporate, competitive, and functional—to ensure that all following activities and tasks that concentrate on mobility are in alignment with where the organization is headed on the grand scheme. To use an example, you wouldn't want to set out to build a Cadillac if your company's general direction is to service a customer segment that consists of 18- to 23-year-old college students. Once your organization's general course is understood and the strategic vision is confirmed as being in alignment, you can map its objectives to the benefits associated with various applications of mobile technologies. Figure 12.4 summarizes some of the technologies' major value propositions.

Next, to make certain your review of your internal organization is comprehensive and doesn't leave any stones unturned, you might want to map your organization along the lines of the all-familiar value chain. Although the value chain concept is slowly being supplanted by the Value Web approach introduced in Chapter 5 (and will also use as part of this analysis), it offers a fast approach to understanding the basic flow of products and information within an enterprise. Lay out the individual links of that chain, all the way from R&D efforts, production, sales and marketing, distribution, to postsale customer service, or whatever the relevant megaprocesses exist within your organization.

Once your organization's value chain is mapped, investigate each process for opportunities to leverage mobile and/or wireless technologies. Look for angles, or hooks, where such functionality might take hold. When investigating enterprise-focused opportunities you probably have a good idea already as to where the bottlenecks occur in your organization, what processes would benefit from being streamlined, or the major cost pools you could diminish by deploying a mobile application. Similarly, when looking into customer-facing opportunities you are probably aware of the major interfaces between your organization and your customer base.

General Benefits from Mobile/Wireless Applications

Increase Revenues Through Increased Sales Opportunities

- Sell additional products and services through the new, wireless sales channel
- Create completely new service offerings that can only be offered via wireless
- Stimulate impulse buying behavior
- Stimulate sales by providing timely information anytime, anywhere (e.g., stock trading)
- Increase yield of capacity services (cinema, transportation, hospitality)
- Increase opportunities for localized targeting
- Increase opportunities for personalization
- Increase opportunities for push marketing techniques
- Increase access to CRM applications to facilitate sales by a remote workforce

Lower Costs Through Greater Productivity and Efficiency

- Eliminate manual labor, which allows staff realignment
- Reduce or eliminate unproductive travel
- Reduce or eliminate redundant data entry (paperwork)
- Improve data accuracy and avoid rework by entering data only once while on location
- Provide instant access to last-minute, critical information
- Improve communication between the company and employees, suppliers, and business partners
- Reduce voice communication costs
- Decrease cost of distribution
- Reduce customer service costs through automated support

Improve Customer Satisfaction

- Provide better, more consistent service due to a more effective remote workforce (sales, customer service, repair/maintenance, etc.)
- Provide an additional, convenient channel for customers on-the-go to do business with you
- Enhance your existing offering via the value-added features of mobility, location awareness, and time sensitivity

Improve Operational Effectiveness

- Provide a means to automate and streamline the remote workforce
- Automate and streamline back-end operations, including systems monitoring, logistics and warehousing
- Provide a means to improve supply chain integration

Source: CGE&Y. J. Teutloff

Figure 12.4 General benefits from mobile/wireless applications.

The most effective way to discover these opportunities is to talk with the individuals who work within each link of the chain. Your colleagues might be able to quickly tell you what you are looking for simply due to their exposure to the daily operations. In addition to being able to point out opportunities, these folks are likely to represent the user communities who have to work with the solution once it's deployed and thus should be involved in the strategy-formulation process.

When researching opportunities for mobile/wireless applications you'll quickly notice two major subgroups—enterprise-external and enterprise-internal applications. External applications frequently are customer-focused, although they can also provide mobile data access to your suppliers or other entities, for example governmental watchdog agencies. In general, external applications provide your customers with another channel to do business with you. These interactions might include mobile commerce, that is a customer buying a product or service from you using a mobile device, or they can center around granting a customer mobile access to data and information, including product specifications, customer service, and the like. On the other hand, enterprise-internal applications provide your employees using mobile and/or wireless devices with access to information residing on the corporate network. Such internal applications include accessing corporate e-mail and calendaring applications, or checking inventory status and price lists.

Besides identifying potential opportunities for mobility applications by scrutinizing each link of your organization's value chain, you need to ask yourself candidly about your company's readiness and capabilities to deliver them. Mobile technology is a very young field in most industries, it's rapidly changing and evolving, which makes it very difficult to attract and retain the resources that can actually build and/or integrate the systems. Instead of training your internal staff, or ramping up recruitment to hire knowledgeable technology resources, it might make more sense to outsource this type of work, especially in the short run. We have found that once a technology matures, bringing specialized resources in house can be an advisable strategy to control costs. Yet when dealing with a moving target, organizations are usually better served by outsourcing the work to companies that have done this type of work multiple times and have assembled a team of experienced mobile technology specialists who continuously sharpen their skills as the technology evolves.

External Assessment

Once your internal analysis is complete, you probably have identified a few areas that would benefit from mobile enhancements. If some of those opportunities involved value chain links that touch entities outside of your enterprise, such as customers, partners, or suppliers, your next steps should be to talk with these parties to solicit their input. Especially when your potential solutions entail functionality to be used by your customers, it is advisable to collect their wants and needs. Engage your customers via focus groups, interviews, surveys, or whichever other means appropriate prior to deployment to ensure these constituents are willing and able to work with the new solution.

Next, your external assessment should certainly consider the mobility efforts others within and outside of your industry are deploying. It's time to leave your own backyard and look at what the neighbors are doing! A review of their Web sites can reveal to you the mobile technologies your competitors have deployed especially in the consumer-facing realm. In addition, search your competitors' press releases to find intelligence about initiatives underway. This is especially true if the company partnered with a point solution provider who, if a new company, is likely to use the project as a reference and success story to generate future work. Having said that, a general search for news releases or editorials about your competitors should be complemented by keyword searches for "wireless" and "mobile" and the competitor's name on meta search engines such as google.com or dogpile.com. Similarly, if you are subscribing to the Dow Jones News Retrieval service, Factiva, or a similar service, you can search hundreds of publications for stories about other companies and their initiatives. Last, third-party research outfits frequently publish reports and case studies that discuss successful—and failed—technology deployments.

Although investigating your peers is a good starting point to learn about mobile technologies that are relevant to your own organization, you also want to widen your search to include companies you are not directly competing with. A broader search would investigate leading mobile/wireless practices in general, regardless of what industry they are deployed in. These applications might be directly transferable to your own organization, and at the very minimum, learning about them should get you started thinking outside the box. It is very likely that other companies in entirely different industries have found the solution to your problems. Just because nobody else in your sector has adopted this solution doesn't mean it shouldn't be included in your initiative portfolio.

Last, but not least, a thorough external analysis includes the assessment of industry trends that are expected to shape the playing field in the years ahead. The reason to look beyond today's realities are obvious; when deploying applications that themselves are just at the brink of moving into the mainstream there might be synergies to be gained that exceed what is attainable with the limited technologies of today. In other words, knowing where your industry is headed can influence your choice of solution. For example, if your industry is becoming increasingly cost competitive, you might want to focus your attention on applications that will allow you to reduce operating expenses. Alternately, if an accelerating pace towards globalization marks your industry, the primary drivers for your mobile/wireless solution might be cross-country data communication standards.

Once you have completed your internal and external assessments, it's time to combine these insights and build your Value Web.

Value Web Analysis

At the conclusion of your external and internal audits you will have enough information to start the creation of a portfolio of mobile/wireless initiatives, yet we suggest one additional exercise prior to moving forward. This additional step is the creation of a Value Web for your organization. If you remember, we previously presented the Value Web as a strategy tool that would allow you to identify the relationships between constituencies within your market space. The value of the network view, we argued, lies in the fact that no one company can provide all the products and services desired by today's customers. Instead of trying to go it alone, we said that astute companies first identify the constituents that contribute to the customer experience life cycle and then attempt to control the construct via strategic alliances in an effort to attain sustainable competitive advantage.

As today's companies face the challenges of a rapidly changing new economy and ever-changing technologies, it is imperative that they capitalize on every feasible opportunity to improve their competitive position. Knowing your organization's position in your immediate Value Web is critical in determining the initiatives that will create a competitive market advantage.

We are not going to repeat the detail presented to you earlier in this book, but instead will list the major steps in creating and analyzing a Value Web to refresh your memory. For a detailed discussion of each step, please refer back to Chapter 5.

1. Define your company's offering and its competitive space.
2. Identify and evaluate the market segments your solution is targeting.
3. Map customer needs and Value Web constituents along the customer experience life cycle.
4. Define the value transactions between each Value Web constituent.
5. Create a value index that quantifies the value each constituent brings and receives, and determine each constituent's importance to the Web via the sustainability index. Then, map both indexes along the power grid.

Once this exercise is complete, you need to evaluate the power grid to identify your place within the network. Being aware of your current position—opportunist, commodity player, contender, or power player—will influence the strategic direction of your mobility endeavors. Combining the insights gained from the Value Web analysis with your internal and external assessments provides you with a rock-solid foundation from which to launch the formulation of a portfolio of initiatives.

Stage 3: Creating an Initiative Portfolio

The next stage in the mobile strategy formulation process is the creation of a portfolio of solution candidates. The portfolio holds all potential solutions that you have deemed relevant to your organization. Usually we see initiative portfolios containing anywhere from 5 to 10 initiatives, all of which must have enough supporting detail behind them to facilitate their ranking against each other. These solution candidates should be described in enough detail for user groups and executive management to be able to prioritize them and decide on their implementation.

Figure 12.5 shows the activities associated with the creation of a set of potential project candidates—specifying each initiative's functionality, technology requirements, costs, and benefits, as well as the impact on the organization.

Functionality Definition

The most obvious descriptor of each initiative is the functionality it brings to the identified user group. What is the mobile solution trying to accomplish? What are the business requirements the solution addresses? How does it address these requirements? What are the features of the solution? How are users going to access this functionality? These are some of the questions that should be answered regardless of whether the initiative embodies an enterprise-internal or external application.

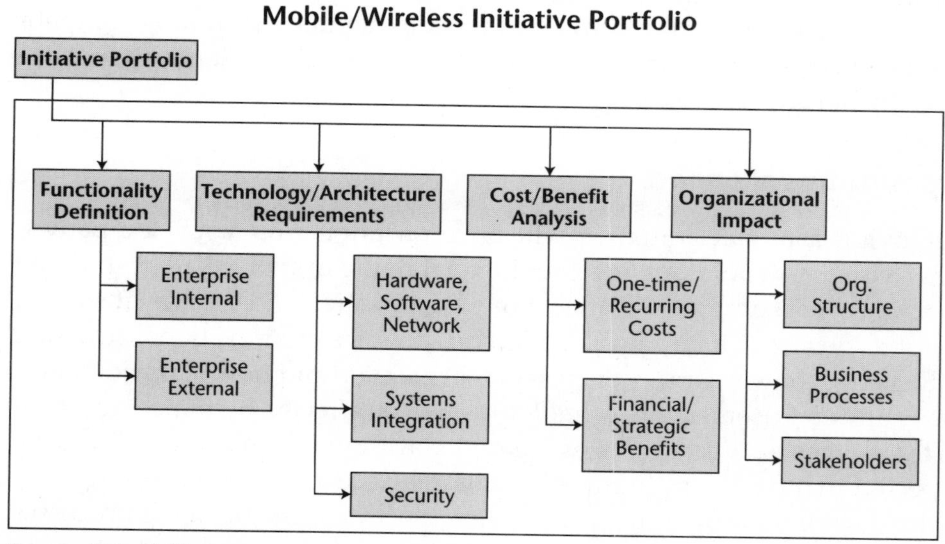

Source: J. Teutloff

Figure 12.5 Mobile/wireless initiative portfolio.

Technology/Architectural Requirements

Next, your discussion of each initiative to be included in your portfolio should address a high-level assessment of the technologies involved and the architecture required to support the solution. What are the architectural components that make the initiative work? What kind of hardware runs the application, and what kind will need to be acquired and distributed to the user groups? How is the solution serviced and maintained? What software needs to be purchased or built? How is the solution going to integrate with the organization's back-end? What types of data transfers need to occur? What's the impact on your network? Are there additional processing or storage requirements? How are you going to address security requirements? You can keep this analysis at a high level for now, because these questions will be answered in detail once a final solution is identified and cleared for deployment.

Cost-Benefit Analysis

Another critical descriptor of each initiative is a high-level cost-benefit analysis. On the cost side, you will want to get a rough understanding of the one-time and recurring costs that are required to deploy the solution. Such costs can include hardware/software, network and security expenditures, and professional fees, as well as expenditures for maintenance, user training, and so forth. On the benefits side of the equation, you should identify the financial ROI to be generated from additional revenue or a reduction in operating expenses. Similarly, there are likely going to be a slew of strategic advantages you will want to mention, including competitive positioning, enhancing your operation's scalability, and other non-quantifiable advantages.

Organizational Impact

Deploying a mobile initiative will have an impact on your company—otherwise you would not have deployed it in the first place. Start thinking about how the solution will affect your organization and the way it's going about its business. If applicable, point out how the organizational structure will change, how business processes will change, and how stakeholders—especially end users of the technology—will be affected by the deployment of the suggested initiative. For example, at the very minimum it can be expected that the user groups who will interface with the technology will require some training. If the technology affects specific business processes you might be facing new workflows due to a reduction in staff that previously performed manual activities—activities that are now automated.

MILESTONE B: INITIATIVE RANKING

Once the portfolio is complete and the individual initiatives are clearly defined, you are ready to assess the initiatives against each other to arrive at a prioritized list to guide your implementation efforts. The activity of raking initiatives is a critical one that warrants the involvement of those stakeholders who are going to be affected by the solutions proposed. As mentioned earlier, to ensure buy-in and to instill a sense of ownership, the users who will be working with the new technology should be involved in the prioritization effort, as should the key decision makers and project sponsors who will oversee and manage any implementation efforts.

The actual ranking exercise is frequently accomplished in an off-site session that brings together decision makers and representatives of the user communities. At CGE&Y client engagements, we usually arrange one or two day-long sessions, during which the initiatives are presented, the team discusses their merits, and agrees on ranking criteria. Depending on your organization, such criteria might include financial resources required, financial return anticipated, development complexity, ease of deployment, strategic benefits anticipated, and others.

Once the ranking criteria are agreed upon, all attendees vote in an anonymous fashion, with each member having the same number of votes to be distributed among the proposed initiative candidates. Then, the votes are tallied. The result is a prioritized initiative portfolio, containing sets of high-priority, medium-priority, and low-priority initiatives.

The set of high-priority initiatives, possibly up to three projects, depending on complexity, include projects that will be cleared for immediate build-out. (This assumes a positive outcome of the go/no-go decision made upon review of a limited-functionality prototype.) Medium-priority initiatives might merit deployment, yet lack the urgency or ROI justification required for immediate consideration. These initiatives are usually placed on the back burner, to be evaluated again after the successful launch of those projects contained in the first set. Low-priority initiatives include projects that have been shelved until a later point in time.

High-Level Deployment Roadmap

Now that you have arrived at a set of high-priority initiatives, you can start thinking about how they will actually be rolled out. Although a high-level deployment roadmap does not specify in detail how you will go about deploying each initiative, it does introduce some of the approaches you should consider prior to implementation. Your high-level deployment roadmap should include your initial advance to building the solution, forming alliances, marketing the solution, and addressing potential challenges and risks that frequently crop up during the implementation phase (see Figure 12.6).

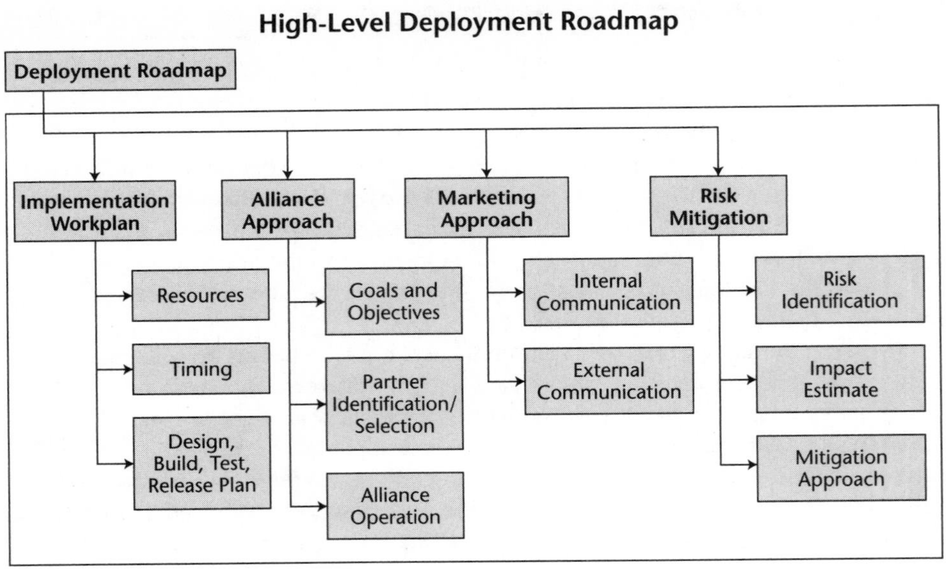

High-Level Deployment Roadmap

Deployment Roadmap

Implementation Workplan
Resources
Timing
Design, Build, Test, Release Plan

Alliance Approach
Goals and Objectives
Partner Identification/ Selection
Alliance Operation

Marketing Approach
Internal Communication
External Communication

Risk Mitigation
Risk Identification
Impact Estimate
Mitigation Approach

Source: J. Teutloff

Figure 12.6 High-level deployment roadmap.

Implementation Workplan

A high-level implementation workplan captures your approach as relating to the actual development and construction of the mobile application. You are going to build a detailed workplan once a proof-of-concept results in a "go" decision, but even before that it is a good idea to map out how you are planning to tackle the challenges associated with building the solution. When preparing a high-level implementation workplan, you should consider the resources that are going to be required, the timing of the process, as well as the activities surrounding the design, build-out, testing, and release planning.

Alliance Approach

A key factor for a successful deployment of a mobile system frequently includes the development of a solid alliance strategy. The task of implementing a technology solution is difficult and can be very complex. Very few companies have the resources and skills to complete the tasks on their own. More importantly, strong business partnerships open a whole world of opportunities with new customers and resources.

No matter whether the high-priority initiative focus on company-internal or -external applications, before taking the first steps in building and deploying your solution(s), you should evaluate the benefits of partnering with other members located within your own Value Web or the general Wireless Value Web presented earlier. For example, such an alliance could involve a preferred hardware and/or software vendor whose technologies you will employ when building your solution. Especially when you are dealing with new companies that have not yet reached mass-market recognition, you might be able to come to an arrangement where the vendor would provide you its technology under preferred conditions, whereas in return you would allow them to cite the project as a reference.

Besides gaining access to new technologies, there are other reasons to partner with companies during the deployment of your initiatives. Such reasons include reaching new customers, accessing new markets, broadening your product/service portfolio, creating economies of scale, enhancing customer service operations, accessing new distribution channels, mitigating risk, and accessing sources of capital. Partnerships with device manufacturers, content providers, connectivity companies, technology, or environmental enablers might help bring that strategy to fruition faster than were you to work on your own.

During your assessment of potential alliances, there are a few steps you should follow. Although a detailed discussion of the alliance approach exceeds the scope of this book, we want to mention them for your consideration. First, you should conduct a thorough assessment of your organization's current state and identify what your organization will look like after the deployment of the selected mobile applications. Second, if you think that aligning yourself with another company might be beneficial to your cause, you need to specify the strategic objectives you are trying to accomplish and the feasibility of reaching them through a combination. Next, if the objectives are clear and the feasibility is real, you will need to identify potential candidates and ensure that their competencies meet your requirements. This task involves the development of relevant screening criteria that are going to be applied to a long list of potential alliance candidates. The result will be a targeted short list of partners that exhibit a strong strategic, cultural, and operational fit. Part of this step should be to clearly understand the economic value each party brings to the table, and the potential impact an alliance will have on both organizations. Now, after completing your due diligence, you are ready to approach each candidate with your value proposition and the appropriate deal structure that might take the form of a joint venture, minority equity, active contractual, or passive contractual partnership. Last, ensure that you regularly monitor progress and track the results of your alliance by establishing the appropriate metrics and controls at the outset to ensure ongoing success of the partnership. Part of this process should be to have a mechanism for making mid-course adjustments to adapt to changing market conditions—which includes having a clear exit strategy.

Marketing Approach

Even before you actually deploy your solution you should get your marketing department involved in crafting the message to be communicated within the organization and/or outside of it. Internal communications should inform your staff about the initiative to be deployed, their benefits, and how they will affect the company as a whole—especially the user groups. If your solution is targeting your customers, suppliers, or business partners, the same idea applies—let everybody know what you are doing so that there are no surprises when the solution is ready to be released. Use the communication to create excitement about the new technology, bearing in mind that not knowing about why you are deploying the solution and how it's used might cause unwarranted and more importantly avoidable anxiety within the user community.

Risk Mitigation

Finally, your deployment roadmap should have a high-level contingency plan in place that identifies the potential risks associated with the deployment of the technology, the anticipated impact that an issue might have on the project and/or the organization, and potential means to mitigate that risk. Let's face it, very few if any technology deployments are completely smooth and flawless. More often than not you will run into challenges—challenges that can be overcome with adequate planning. Preparing a risk mitigation approach at the very minimum makes you think about what could go wrong during the project, and ideally results in a tangible plan of action to face the challenges associates with the technology deployment and potential avenues to protect yourself against major implications.

Creating the Proof of Concept

Before the go/no-go decision is made—usually leading to the commitment of substantial financial and other resources to the build-out of an initiative—a proof of concept, or pilot, should be created. Used as one last point of evaluation before the construction of the full-blown solution, the prototype could be a rough rendition of what the final solution might look like. Similarly, when talking about an installation of a wireless LAN, for example, a pilot deployment would entail a limited roll-out of a solution to a small sample of the targeted user community (see Figure 12.7).

The purpose of the proof of concept is to allow the users and initiative sponsors to evaluate and assess the mobile/wireless solution on a small scale and with limited functionality prior to launching full-scale implementation efforts. The proof of concept is a low-risk, low-resource demonstration and validation of the final solution. Proof-of-concept applications are usually throwaways that are discarded after evaluation. The final solution is usually built from scratch; rarely do prototypes serve as the platform upon which that solution is built. On the other hand, we have deployed several wireless LAN pilots that were being used as "beachheads" for larger deployments. Once a WLAN is in place, it is relatively easy to expand the network and widen its reach—no need to start all over.

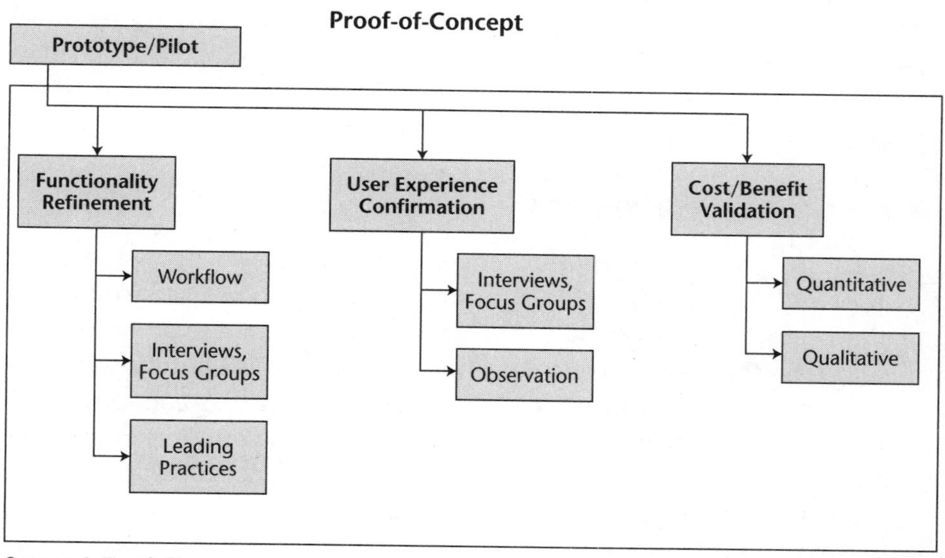

Source: J. Teutloff

Figure 12.7 A proof of concept is a final checkpoint used to validate expectations.

Functionality Refinement

Deploying a limited-scale prototype or pilot offers the team a great opportunity to refine the solution prior to a massive-scale roll-out. Proof-of-concept applications might illustrate some of the solution's functionality by incorporating some of the features to be integrated into the full-blown application. For example, the prototype might link into your back-end system, being able to pull *out* data and display it on the mobile device, yet the limited nature of the application might prevent the data transfer *into* any legacy applications. Similarly, a limited pilot deployment of a wireless LAN, for example, might include a small group of users in one single location, maybe one of the company's various departments, or one select pilot branch office.

The goal of the proof of concept, of course, is the verification and refinement of the system's functionality. As mentioned earlier, not only can mobile technologies streamline traditional work processes, but they can also spawn entirely new workflows. The proof of concept will allow the solution development team to refine features by investigating and incorporating such new workflows, by interviewing the participants in the pilot project and learning from their comments, and by applying leading practices that have been observed at other companies that have successfully deployed a similar mobile solution.

User Experience Confirmation

In addition to giving the solution development team an opportunity to refine the solution's functionality, a proof of concept can be used to confirm and enhance the user experience. At the very minimum, a prototype of an application is a static series of screens that shows what the solution should look like. These screens commonly illustrate the overall creative design, screen layout, color scheme, and navigational flow. The intent is to provide a visual rendition of the application and its user interface while at the same time offering a glimpse into what type of data and/or information would be presented and in which format. The focus of illustrating the user experience via a proof of concept is to demonstrate the solution's ease of usability. The solution development team might conduct focus groups, in-depth interviews, surveys, or observations among the user community to confirm the usability of the solution and to tweak it prior to finalization.

MILESTONE C: THE GO/NO-GO DECISION

Once the prototype confirms the functionality, usability, and ROI of the mobile solution, the executive sponsors make the final go/no-go decision. A positive decision secures the second round of funding required for the construction of your initiative. Once the go decision is made, your solution development team—whether it's recruited internally or outsourced to a systems integrator—shifts into high gear and starts the full-blown effort of Phase II—strategy implementation.

Cost-Benefit Validation

Last, but certainly not least, the proof of concept is an important event in the mobile solution development process to verify the system's anticipated qualitative and quantitative costs and benefits. Frequently, we observe that cost estimates applied during the planning stages need to be revised, or that the solution provided benefits previously not anticipated. In other words, the pendulum can swing either way once the rubber hits the road. And that's exactly the purpose of the limited deployment; validation of the planned expense or capital expenditure from a perspective of financial and strategic ROI.

Phase II: Strategy Implementation

Whereas Phase I was concerned with the formulation of your organization's mobility direction and the development of specific solutions that best correspond to internal opportunities and capabilities in light of your organization's external environment and leading technology practices, Phase II is where everything comes together. Your various planning activities during Phase I have provided you with high-level approaches that now must be refined to guide your ensuing implementation efforts.

Because the methodology of actually building a mobile solution is very complex and takes up an entire book in itself, Chapter 13 presents selected excerpts from *The Cap Gemini Ernst & Young Guide to Wireless Enterprise Application Architecture*. This previously published book goes into the intricate, technological details of building a mobile infrastructure, and thus serves as a good companion to this volume's focus on the strategic business value provided by mobile technologies.

Phase III: Strategy Monitoring

After developing and deploying your solution, Phase III entails a rigorous monitoring of what you have created. To ensure you are reaching your business objectives set out during Phase I, you should keep close tabs on the benefits that were anticipated and that are actually achieved. If you are not reaching your goals due to challenges that you did not anticipate and account for, be prepared to pull the plug and go back to the drawing board.

Adjusting your course after deployment is a twofold exercise. First, evaluate the lower-level priority initiatives that you formulated but placed on the back burner during your prioritization of your mobile/wireless initiative portfolio. Conditions might have changed to the point where those solutions are now a higher priority. Second, your internal and external environments are likely to have changed from the time you performed your initial assessments. In other words, you might need to go back and once again survey your internal business requirements and capabilities and map those against technology solutions deployed within or outside of your industry.

At the minimum, you will find that technology has not stood still, nor have your competitors. Further, other entities outside of your industry might have developed and deployed mobile solutions that could be leveraged in your business environment. A thorough external audit should be performed on a regular basis, regardless of whether you are changing your current strategy and initiatives. Strategy formulation is a continuous undertaking, not a once-a-year activity to be conducted and then shelved for the next 12 months.

Summary

This concludes the illustration of the high-level process applied to deploying mobile technology solutions within an organization. To summarize, before jumping into implementing any technology, you should make sure you follow a structured approach in an effort to avoid—or at least minimize—costly missteps. The three phases of developing a solution entail formulating your mobile/wireless strategy, implementing that strategy and monitoring success. Phase I starts with a shared vision that, once funded, leads to the creation of a mobility direction for your company. This direction will reflect a candid assessment of your business requirements and capabilities, external opportunities and leading practices, and a thorough Value Web analysis. A clear direction is the foundation for the creation of a portfolio of mobile technology initiatives including high-level functionality definitions, technology requirements, cost-benefit analyses, and assessment of the initiatives' impact on your organization.

After ranking the initiatives in a collaborative exercise to ensure buy-in and support from both executive leadership and user communities, you are ready to draft a high-level deployment roadmap for your top-priority projects. This roadmap includes an initial work plan, an approach to forming alliances, marketing considerations, and thoughts around risk mitigation. Next comes the creation of a proof of concept that validates the functionality, user experience, and anticipated ROI. The subsequent go/no-go decision seals the fate on your proposed application. If the decision is a positive one, your project moves into Phase II: strategy implementation, the topic of the next chapter.

CHAPTER

13

Methodology for Building the Solution

Adam Kornak

Until now, the majority of the topics in this book have been around the business/strategic aspects of mobility and wireless technologies. This last chapter switches gears slightly and covers the leading practices around the methods of building an enterprise-wide solution as well as some of the pitfalls that you may experience when building a mobile application.

One of the first things to understand when you're considering building a mobile application in your organization is to start small, perhaps with a pilot implementation. Many organizations that are considering integrating mobility applications jump too quickly into the mobile space for fear of missing out on the benefits. The result, quite often, entails dissatisfied or confused customers, a misdirected mobile strategy or lack of a strategy, and ultimately an expensive application that is put on the shelf and never used. In the end, you might find yourself with an investment that provides little or no return—a situation most of us would like to avoid.

Another common scenario we have observed entails that many organizations are led to believe that supporting these new applications requires an entire rebuild of the existing IT infrastructure. In reality, mobility is just an extension of your current enterprise architecture and should be treated as such when designing new mobile applications. Of course, that doesn't mean that implementing a mobile application is a simple process that can be accomplished without applying the usual rigorous software development standards. Even more importantly, organizations tend to forget the business reason for taking the path down mobility. There are many exciting aspects of mobility that make it very easy to lose sight of why you wanted to implement in the first place! The dynamics of mobility are simply mind-boggling. It seems that every time we turn around a new device or application has entered the marketplace. The excitement of owning the next sleek and powerful gadget is hard to relinquish.

This chapter seeks to help you understand the various concepts of software development used to establish a framework for building mobile solutions. In addition, the intent of this chapter is to help you understand the basic concepts of the methodologies that were used in the case studies in this book and throughout our experiences in mobile systems integration projects. This chapter does *not* cover all the architectural design components that you'll need to understand when designing a mobile strategy, but it *does* provide you with a baseline understanding of the design principles of enterprise architecture that CGE&Y has successfully used in the past. Let's get started with a discussion on the adaptive nature of architecture and methodologies.

Adaptive Technology Architecture Definition (ATAD)

Technology architecture is different in an adaptive IT organization—the architecture must be able to accommodate the rapid introduction of new services and the decommissioning of old ones as the needs of the organization change. The architecture must take into account not only the well-defined technology elements within the company but also technology elements provided by partners across the ecosystem—elements that may be defined only by their interface or by the service they provide. Above all, the architecture must be easy to modify so that it remains a relevant planning tool under rapidly changing IT needs. These concepts apply regardless of whether the application is wireless or wired.

Organizations that fail to adequately utilize technology will not achieve its full potential. Conversely, technology that is not in alignment with the governing business objectives will fail to adequately support the enterprise. The primary goal in any architecture and methodology, then, is to ensure that your operation and technology are in alignment. Characteristics of a solid methodology include the following:

- **Value focused**. A market-leading and consistent perspective on architecture is focused on delivering value to your clients and organization. A value-based characteristic forms the link with your business and IT organization. The technique of *roadmapping* (a method of architecture focused on mapping the layers of architecture) ensures focus on market demand, taking situational factors into account in order to realize optimal value add.
- **Scalable**. Architecture should be scalable in order to provide pragmatic guidance to the IT community at the enterprise and at project level. You also need to be able to adapt to the situation and environment. Therefore, it's important to discern different levels of architecture content (such as enterprise- and project-level architectures) and architecture aspect areas, easily covered within full integration or partial alignment.
- **Flexible.** It's important to form a flexible, efficient, and stable approach for delivering architecture support to projects and standalone architectural services. The approach should not contain a *prescriptive methodology* (in others words, a standard step-by step approach), but instead should form a stable platform for innovation. Architecture separates process from content, both addressed within a framework. The concept of roadmapping is the key asset that links process and content in order to effectively meet customer demands, required business issues, and engagement efficiency.
- **Integrated.** As a market-leading discipline, architecture integrates the full scope of business and technology issues. The CGE&Y Architecture Framework defines architectural issues and their relations and interdependencies according to different levels of contemplation. For more detailed information, please see the *CGE&Y Guide to Wireless Enterprise Application Architecture.* (Wiley Publishing, Inc.).
- **Common mindset**. A successful architecture toolset provides a common mindset and language for your architecting community and your clients. It needs to be as simple as possible to enhance transferability. Thus, a solid architecture terminology should be defined, and best practices integrated within the approach.

- **Solid foundation.** Architecture should provide support for the process and content of all architecture-related services. Architecture frameworks describe critical outputs and capture successful situation-appropriate approaches for their development. Defining the scope of architecture work and formalizing the roles and responsibilities are critical tasks of this characteristic.
- **Fast and efficient.** An architectural approach should encourage the reuse of innovative or proven practices, thus greatly facilitating and accelerating the development of client solutions. The approach encloses application of techniques such as patterns, roadmaps, separation of concerns, and usage of tools.

In CGE&Y's architecture methods, an element known as the pyramid of IT needs addresses all levels of a standard hierarchy of needs (see Figure 13.1).

Only on top of a solid technical infrastructure and security architecture can you build successful information systems with their specific security requirements. And only when you are successful in maintaining these applications and keep the environment secure will the corporation and/or organization be successful realizing its business strategies. By achieving this basic pyramid of needs requirements, the IT area will then be recognized as the enabler of business and business changes.

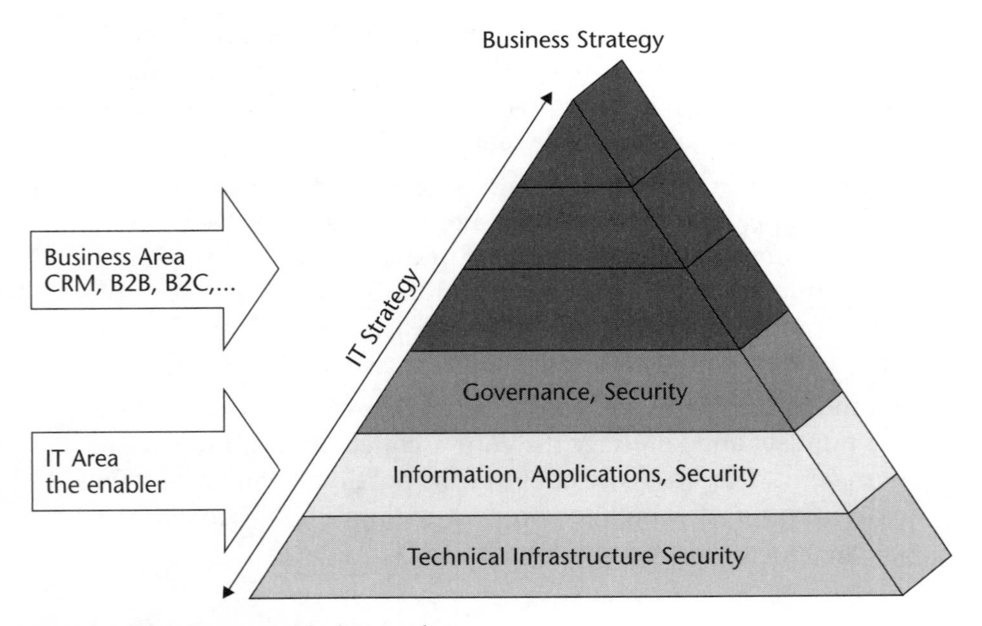

Figure 13.1 The pyramid of IT needs.

The CGE&Y architectural methods follow a layered abstraction approach to the complex issue of defining how technology will support business needs. This multiphase abstraction is critical to the process of deriving technology requirements from business strategy (see Figure 13.2).

The first phase of this process is called the *contextual phase*. You initiate the project by gathering information to fully support the business case and answer the question "Why we are doing this?" Gathering the necessary information to validate the full scope of the project and to generate the detailed project plans used going forward are completed during this important phase.

In the *conceptual phase*, you articulate the business strategy in terms that will allow you to determine exactly what you will do from a business perspective so that the architecture will reflect and support those requirements. During this phase, all distraction factors—including organizational boundaries, political issues, technology biases, and geography—are broken out in order to obtain a simple, clear picture of what the architecture has to achieve.

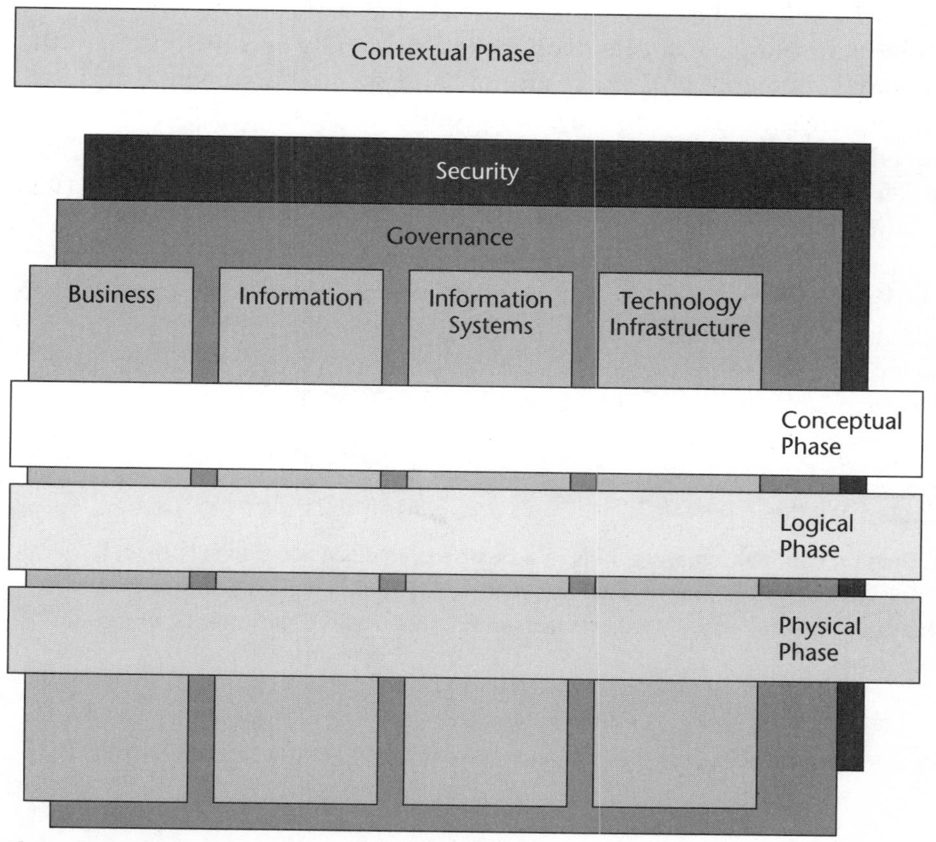

Figure 13.2 Overview and layers of architecture.

In the *logical phase*, you address how the business strategy can be realized. Scenario planning is utilized in order to determine the best solution and to resolve business imperatives that might conflict ("we want the lowest risk solution with the highest flexibility," for example). At this stage, you begin to reintroduce complexity factors such as geography and organizational boundaries. At the end of the logical phase, the optimal scenario is determined and taken forward into the next phase.

In the physical phase, you determine what tools and resources can be used to create the solution. At this point, you identify actual products for services that will be utilized to implement the architecture. You also address distribution of data and business logic as well as the physical implementation of platforms (for example, one or multiple servers? clustering or failover?).

The Purpose of Architecture

Architecture describes overarching designs of individual components so that their assembly results in a complete and working product. This design is needed to guide the construction and assembly of components.

Architecture was, is, and will always be driven by specific needs. In the IT arena, architecture will always follow and specify incremental and iterative implementations. The business needs drive for speed to market and reduced costs, which start earning their ROI right from the beginning. This implies phased implementation principles such as the famous 80:20 rule. Without a concise and precise blueprint or enterprise architecture there is no measure in determining which 80 percent to realize and which 20 percent to skip, at least initially. The level of complexity in today's application landscapes is prohibitive to well educated decisions incorporating the most critical dependencies without an existing enterprise architecture.

During each architecture project the architect(s) will have to build a framework for domains.

NOTE What is a Domain?

A domain can mean many things in a business/IT discussion. In this case, it refers to a separate and distinct business entity that is logically subdivided, for example, into organizational business units, functional departments, or market segments.

These domains will be created based on criteria derived out of the business and/or IT principles. The criteria can be function-like, such as geographies, business units, levels of security, and so on. But what all domains have in common is the collection of IT services they have to provide in order to support the business process flows. It is tedious but very rewarding work to identify duplicate services throughout a corporation's domains. Very often activities such as these are driven by data ownership principles. Owner of data entities will have to provide all services needed by all other domains. Clearly, even in an ideal environment, new developments are not simply rearrangements of existing components to form new applications. There will always be some individual coding necessary. But the utilization of reusable components will reduce project durations and costs and, as a side effect, enforce standards. Additional accelerators are roadmaps, describing how to approach the development of a solution, potential hurdles, checklists, and so on, and patterns, describing partial solutions to specific parts of the solution, very similar to the plug-and-play approach in other IT industries.

Aspect Areas

Just as a solution is built from multiple components, a balanced architecture will have to reflect more than just one aspect of a solution. An aspect area can be viewed as a view from one specific viewpoint at the solution with one specific objective. Each objective results in a set of related structuring criteria.

The topics addressed in each of the aspect areas differ widely, and the complexity in each of these areas requires that specialists be engaged for each area. However, there are strong interdependencies between each of the aspect areas, because the business structure prescribes information structure, which in turn prescribes information systems structure, which then prescribes technology infrastructure. Thus, ideally all aspect areas have to be incorporated into the architecture design to ensure its usability as one system. Skipping one of the aspect areas implies additional risk, because critical viewpoints have not been considered. Designing solutions based on individual aspect criteria reduces complexity and creates clarity of mind. Only when all aspect areas are overlaid and create the complete solution do you add complexities and reflect a realistic but now balanced view of the solution. Table 13.1 defines the content behind each of these aspect areas.

Table 13.1 Sample Aspect Areas of Architecture

ASPECT AREA	CONTENT
Business	Noncommercial organization
Human resources	
IT processes	
Marketing	
Information	Information structures
Program management	
Business intelligence	
Processing structure	
Information Systems	Automated IS support
IS services	
Integration of IS services	
Collaboration between services and components	
Technical infrastructure	
Infrastructure support required	
IT services required	
Processing platforms and volumes	
Network designs	
Hardware/software standards and requirements	
Governance and security services per platform (failover, load balancing, encryption, and so on)	
Communication methods and standards	
Governance	Ongoing support of the different business processes
Issues management	
Backup	
Disaster recovery	
Performance	
Security	Business security needs at the different levels of technology platforms, applications, and network infrastructures
Security prevention services and elements, such as single-sign-on and nonrepudiation	

Specialized Views

Developments in business and technology often require special attention in the architectural design. Within the architecture, this translates into the flexibility of integrating any necessary view into the architecture. These views typically don't appear by themselves in the picture. They are selected and derived from recorded business and IT principles. Only a solution that can be measured against evaluation criteria will, in the long term, be successful. Using views and viewpoints supports this evaluation process.

Viewpoint: A pattern or template from which to construct individual views.

View: A representation of a whole system from the perspective of a related set of concerns.

The following is a list of potential candidates for specialized views. This is not an all-inclusive list, but simply intended to provide examples.

- Component view
- Actor view
- Governance view
- Security view
- Data view
- Technical infrastructure view
- External communication view
- Internal communication view
- Data storage view
- Data access view
- Business unit view

Levels of Contemplation

CGE&Y's architectural approach differentiates itself by presenting several different aspect areas for architecture. These aspect areas, although focusing on different elements of the architecture as a whole, still maintain a high degree of cross-communication and integration both internally and with CGE&Y's estimating, delivery, and quality-assurance (QA) methods. This technique is further differentiated because it is supported by a set of integrated methods that cover the various aspect areas of architecture and drive the entire process from conceptual design to detailed physical design and technology selection.

These methods provide a high degree of specialization within architecture aspect areas, while still maintaining the business-driven and integrated approach of the original standard (see Figure 13.3).

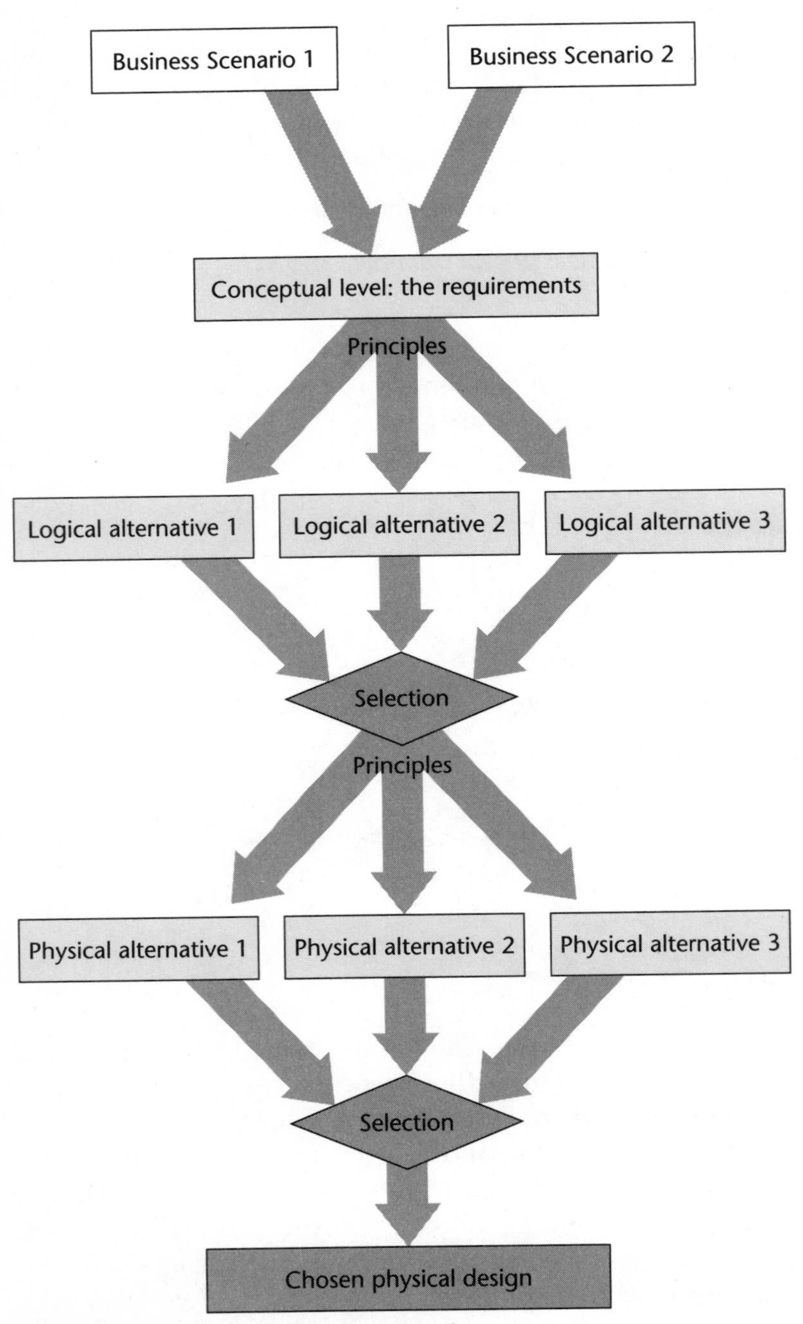

Figure 13.3 A business-driven approach.

The CGE&Y architecture methods define a layered abstraction approach to the complex issue of defining how technology will support business needs. This multiphase abstraction is critical to the process of decomposing technology requirements from business strategy.

As mentioned previously, the first phase of this process is called the *contextual phase*. You initiate the project by gathering information to fully support the business case and determine why your organization is initiating the project. The objective of this phase is to prepare everything needed to start the project. In this phase, there will be three primary sets of deliverables:

- The *Project Quality Plan*, which contains all project-related deliverables necessary for such an architecture project
- The *Internal Quality Plan*, which contains all of your organization's internal management and QA documentation
- The *Architectural Input File*, which contains all relevant input information for the project

Additionally, and very importantly, you need to achieve four objectives during the contextual phase:

- Gather sufficient information regarding the client's business strategy and drivers to identify what will drive, govern, and constrain the architecture.
- Define the desired end state for the use of the technology.
- Set the scope for the overall architecture and further define the effort required to achieve the desired end state.
- Establish agreement on the common set of services to be created and estimated.

The deliverables from this phase include capturing the strategic vision, business requirements, and project goals. These will drive the new architecture and an IT framework for the client's desired environment. Additionally, because the scope of the architecture is not to redefine the information system services needed by the client, but rather to define how they will exist in the new environment defined by the architecture, the existing IS services will be captured and agreed on to enable working with a common set across the different streams.

In the conceptual phase, you articulate the business strategy in terms that will allow you to determine exactly what the organization will do from a business perspective so that the architecture will reflect and support those requirements. During this phase, all distraction factors, including organizational boundaries, political issues, technology biases, and geography are abstracted out in order to obtain a simple, clear picture of what the architecture has to achieve.

The conceptual cycle identifies, structures, and defines the scope of the total architecture. The main objectives of this phase are to:

- Identify all the internal and external elements that must be considered in the architecture. These include external influences such as regulation, emerging industry standards, and market pressures, and internal influences including corporate biases, management directives, member constraints, and other elements.
- Define the entities in the architecture that will support the business requirements and the relationships between them.
- Verify that the model supports the business strategy and conforms to the constraints placed on the architecture by both internal and external factors.

The creation of a product always follows a specific set of tasks and activities. First, the requirements for the product are gathered and structured. Then, the specifications for the product are developed. Finally, the detailed production instructions are designed. Consequently, the conceptual cycle in architecture is most concerned with the structuring of the requirements to produce a concept of the architecture.

The aim of the conceptual phase of the architecture is to create the foundation for all architectural decisions. The conceptual architecture must define the business-level requirements that will justify future investments in technology. In order to develop the framework that will define how technology is used to support the business vision, it is important to remember that all aspect areas of the architecture must be addressed.

In the logical phase, you address how the project can be realized. Scenario planning is utilized in order to determine the best solution and to resolve business imperatives that might conflict ("we want the lowest risk solution with the highest flexibility," for example). At this stage, you begin to reintroduce complexity factors such as geography and organizational boundaries. At the end of the logical phase, the optimal scenario is determined and taken forward into the next phase.

The scope and specific deliverables of the logical and physical architectures to be conducted will be determined at the conclusion of the conceptual cycle. Although the process model remains the same, the content must necessarily differ depending on the specialized focus areas that are most important to the client's business.

During this phase of the architecture, an additional tool is used to visualize and validate the current progress. Two concepts of the future standard IEEE P1471 have been adopted and incorporated the CGE&Y methodology. We use the terms *view* and viewpoint within the method.

The IEEE definition of these terms is:

- **Viewpoint:** a pattern or template from which to construct individual views.
- **View:** a representation of a whole system from the perspective of a related set of concerns.

In other words, you can identify viewpoints and create their views to show the structure from a specific perspective. Examples include an information view, an integration view, and a security view.

The logical cycle will use the output of the conceptual phase and address the following issues:

- All platform services (hardware, operating systems, and so on)
- All networking services (including LANs, WANs, Inter-/intra-/extranets)
- All middleware services (including OLTP monitors, object brokers, and messaging services)
- All common services (databases, authentication, authorization, and audit)
- All shared services (including directories and common desktop applications)
- Logical application component models
- Component logical life cycle
- The best structuring of shared application services across the enterprise

The main objective of this cycle is to translate the conceptual architecture into the (logical) specifications of the architecture. This is accomplished by doing the following:

- Defining the information system components by decomposing the requirements
- Projecting the defined information system components onto the desired logical technological framework

- Creating the specifications of the content of the technical components
- Verifying that the proposed architecture meets the requirements

The objectives for the logical cycle will be achieved by performing the following four tasks:

- Creating the logical architecture outline to define the objectives and constraints
- Studying alternative scenarios and selecting preferred solution(s)
- Modeling the (logical) architecture and verifying its feasibility
- Making a final selection of viable scenarios and presenting these to the principle client for review and verification

The outcome of applying CGE&Y's method is a design for all elements of the data processing support and applications infrastructure along with the architecture that support the critical focus areas of the business. However, the method does not address the detailed functional requirements and specifications of applications that need to be supported by the architecture.

Logical Scenarios

Logical scenarios are constructed based on various business drivers, which might have conflicting objectives. Different scenarios are constructed to emphasize particular architectural drivers, which are weighted according to their importance to the client's business. The client and CGE&Y will derive the best scenario from the various options available and this scenario will be enhanced and detailed so that it can be used as a basis for the physical cycle.

For example, a logical scenario might be driven by least cost. Another might be driven by quickest time to market or be driven by minimum technology risk.

In the physical phase, you determine with what can the project be created. At this point, you identify actual products and services that will be utilized to implement the architecture. We also address distribution of data and business logic as well as the physical implementation of platforms (one or multiple servers? clustering or fail over?).

Development of the technical architecture in the absence of the business architecture that it is required to support will result in "development in a vacuum." This virtually guarantees that the information systems framework that is developed will not be supportive of the business, nor will it be able to feed new business opportunities into the organization when those are presented by innovative use of new technology. To enable this alignment and support it is necessary to do the following:

- Define what the current technology environment is
- Define which business requirements will be supported by the architecture
- Define which business and technology vision and strategies will shape the architecture
- Define which high-level services are required to support the business requirements

By bringing the technology and business focus together, you can ensure that technology is utilized within the client's environment in a planned, coherent way that supports the client's business direction. With this understanding, the client will then be armed with sufficient information to clearly evaluate their architectural alternatives.

After all requirements have been gathered in the conceptual cycle and have been translated into the logical architecture design in the logical cycle, the detailed physical design of the architecture has to be developed. Each detail of every component in the architecture is designed and described. Verification of the usability of the design is carried out by analyzing the required performance, capacity, and throughput in detail and calculating, or testing via benchmarking and/or prototyping, the physical possibilities of the architecture and comparing the results to the requirements.

The physical cycle consists of the same four steps as the logical cycle:

- The cycle is prepared by adding the component data flows (information layer) to the information system components (information systems layer) list to create a total list of components that will use computer resources.
- Alternative scenarios are studied and selected; alternative scenarios will be based on the different technical possibilities.
- The physical architecture is created and analyzed.
- The final solution is selected and presented to the client.

Physical Scenarios

Physical scenarios explore the various options available to deliver the services based on the client's business constraints and the service levels required of the various components. As an alternative and accelerator to this scenario discussion, CGE&Y offers technology-focused workshops during which the client-specific technical environment is defined and described. These workshops are highly intensive and require the presence of all decision makers and stakeholders.

Migration Strategy

Consistent with the integrated approach to developing the solution architecture, in the physical phase you must include the necessary tasks for designing the migration plan. The migration strategy is concerned with communicating with the business to agree on an outlined plan for migrating to the physical architecture. This takes into account ongoing initiatives, business areas that urgently require technology enablement, and interface strategies to allow business operation during the migration period. Using scenarios and views, you can compare the various features of the migration options, evaluate the pros and cons, and then accurately determine the most appropriate approach to move forward. The objective is to define possible scenarios for implementing the architecture. The scenarios defined here should reflect the business drivers, organizational priorities, and financial considerations of the client. The differences in the approach of the migration scenarios can have a profound impact on the organization as a whole; therefore, care should be taken to define and validate only appropriate scenarios. Additionally, where possible, the migration strategy should include timelines and cost estimates for the implementation of the architecture.

Security Strategy

The security strategy defines how security will be implemented and managed at the client, including the identification of required policies, procedures, and organizational entities.

The aim of the security architecture method is to enable architectures that address business security needs at the different levels of technology platforms, applications, and network infrastructures.

The purpose of the security architecture is to provide a framework to lay out a security architecture that is logically and uniformly derived from business drivers. The emphasis is on what level of security is needed, how to realize it, and with which techniques and products to use to implement it. The objective of this approach is to enable architectures that address business security needs at the different levels of technology platforms, applications, and network infrastructures. Advantages of this approach include:

- It provides justification for security expenditures.
- By clearly defining the objectives of the security environment, the validity of a security budget can be assessed.
- By defining the overall security architecture, the risks of an insecure "white spot" are minimized. If there are any white spots, at least they are known.
- Security measures in line with business drivers link business drivers to the security environment. The security environment is therefore easily adaptable to future business changes. When the business changes, the security can change with it.
- It decouples functionality and products.

This standard approach detailed here translates to the conceptual, logical, and physical cycles of the architecture method.

Governance Strategy

A governance architecture is aimed toward deploying information technology to support the entire business. This overall enterprise architecture has to provide a structure for guiding the full life cycle of information systems.

Governance is a broad term used to describe more than just the software-engineering aspect of the architecture. It should be thought of as the concept that addresses the following: daily operations, operational impact analysis, budget management, and interdepartmental interaction behavior, among other things. Identifying and implementing this type of strategy enables the business to manage the enablers such as personnel, capital, housing, and IT on which the business is depending. For all these services, the business wants to control the quality in terms of availability, performance, reliability, continuity, and so on. This strategy defines how technology will be managed at the client, including the identification of required policies, procedures, and organizational entities.

There are four key aspects that need to be implemented in order to realize governance effectively:

- The business needs must be agreed upon in terms of quality of service attributes.
- The delivery of quality of service attributes must be designed and implemented.

- Control processes need to be implemented to check the agreed quality of service attributes against the way they are actually delivered.
- Control processes need to be implemented to check whether business needs are still correctly represented by the defined quality of service attributes.

Software Best Practices

Today, society is extremely dependent on software. From our automobiles, which are becoming more "intelligent" to simple things like our phones, software dominates our daily lives.

The good news is that the process of developing software is slowly improving. The bad news is that this industry is still in its infancy, compared to most other manufacturing industries, which can draw from a much longer history.

Still too many software projects fail for a long list of reasons. Some of the pitfalls of these kinds of projects include:

- Inaccurate understanding of user needs
- Inability to deal with changing requirements
- Modules that don't fit together
- Software that's hard to maintain or extend
- Late discovery of serious project flaws
- Ambiguous and imprecise communication
- Overwhelming complexity
- Undetected inconsistencies in requirements, design, and implementation
- Subjective assessment of project status.

This is the place where best practices come into play. Commercially proven approaches to software development that, when used in combination, will deal with most of the issues of software development and lead to the successful completion of projects.

These best practices are as follows:

- Develop software iteratively
- Manage requirements
- Use components-based architecture

- Visually model software
- Continuously verify software quality
- Control changes to software

Putting It All Together

As illustrated in this chapter, the heart of building a mobile solution lies in crafting a solution that is right for each unique situation. The building blocks to designing a mobile solution are not in the cutting-edge technologies or in the capabilities of software-development tools. Mobility is simply an enabler to a business problem of faster and less expensive ways of increasing productivity, increasing customer satisfaction, reducing expenses, and increasing ROI (return on investment), just to name a few. There have been volumes written on the topic of software methodologies and architecture definitions. Hopefully, this chapter gives you a starting point for developing mobile applications in your business situation. For more detailed information on wireless design and architecture principles, see *The CGE&Y Guide to Wireless Enterprise Application Architecture* (Wiley Publishing, Inc.).

APPENDIX

A

Industry Associations

This appendix lists trade associations, standards organizations, and regulators, which illustrate the wide range of entities in the environmental enablers segment. Please note that the description of each entity was taken from each organization's Web site. Some of these organizations are governmental bodies, some were founded by independent third parties, yet others were launched by commercial enterprises. Again, this list is not exhaustive, and we do not endorse these organizations and/or their activities.

ATIS (Alliance for Telecommunications Industry Solutions): A membership organization that provides the tools necessary for the industry to identify standards, guidelines, and operating procedures that make the interoperability of existing and emerging telecommunications products and services possible.

ANSI (American National Standards Institute): A private, nonprofit organization that administers and coordinates the U.S. voluntary standardization and conformity assessment system. The Institute's mission is to enhance both the global competitiveness of U.S. businesses and the U.S. quality of life by promoting and facilitating voluntary consensus standards and conformity assessment systems, and safeguarding their integrity.

Bluetooth SIG: A special interest group that's primarily a volunteer organization run by employees from member companies. These individuals support a number of working groups that focus on specific areas, such as engineering, qualification, and marketing.

CDMA Development Group: An international consortium of companies who have joined together to lead the adoption and evolution of CDMA (Code Division Multiple Access) wireless systems around the world.

CTIA (Cellular Telecommunications & Internet Association): An international organization that represents all elements of wireless communication—cellular, personal communication services enhanced specialized mobile radio, and mobile satellite services—serving the interests of service providers, manufacturers, and others.

ETSI (European Telecommunications Standards Institute): A not-for-profit organization whose mission is to produce the telecommunications standards that will be used for decades to come throughout Europe and beyond.

FCC (Federal Communications Commission): An independent U.S. government agency charged with regulating interstate and international communications by radio, television, wire, satellite, and cable.

GSM Association (Global System for Mobile Communications): An organization concerned with making wireless work globally.

HRFWG (HomeRF Working Group Inc.): Formed to establish the mass deployment of interoperable wireless networking access devices to both local content and the Internet for voice, data, and streaming media in consumer environments.

IEEE (Institute of Electrical and Electronics Engineers): A nonprofit, technical professional association covering technical areas ranging from computer engineering, biomedical technology, and telecommunications, to electric power, aerospace, and consumer electronics, among others.

ISO (International Organization for Standardization): A worldwide federation of national standards bodies from some 140 countries, one from each country. The mission of ISO is to promote the development of standardization and related activities in the world with a view to facilitating the international exchange of goods and services, and to developing cooperation in the spheres of intellectual, scientific, technological, and economic activity.

ITU (International Telecommunication Union): An international organization within the United Nations system where governments and the private sector coordinate global telecom networks and services.

MAC (Mobile Advisory Council): An advocacy group for mobile computing standards and design.

MWIF (Mobile Wireless Internet Forum): An international nonprofit industry association created to drive acceptance and adoption of a single open mobile wireless and Internet architecture that is independent of the access technology.

NIST (National Institute of Standards and Technology): A nonregulatory federal agency within the U.S. Commerce Department's Technology Administration. NIST's mission is to develop and promote measurements, standards, and technology to enhance productivity, facilitate trade, and improve the quality of life.

NTIA (National Telecommunications and Information Administration): An agency of the U.S. Department of Commerce and the Executive Branch's principal voice on domestic and international telecommunications and information technology issues.

RA (Radiocommunications Agency): An executive agency of the United Kingdom Department of Trade and Industry, responsible for the management of the nonmilitary radio spectrum in the U.K., which involves international representation, commissioning research, allocating spectrum and licensing its use, as well as keeping the radio spectrum clean.

TIA (Telecommunications Industry Association): A trade association serving the communications and information technology industry, with proven strengths in market development, trade shows, domestic and international advocacy, standards development, and enabling e-business.

UMTS (Universal Mobile Telecommunications System) Forum: A nonprofit, cross-industry organization that is uniquely committed to the successful introduction and development of UMTS/IMT-2000 *third generation* mobile communications systems.

UWCC (Universal Wireless Communications Consortium): An international consortium of more than 100 wireless carriers and vendors supporting the TDMA, EDGE, and WIN technology standards and their interoperability with GSM and UMTS.

WAP Forum: An industry association that has developed the de facto world standard for wireless information and telephony services on digital mobile phones and other wireless terminals. The primary goal of the WAP Forum is to bring together companies from all segments of the wireless industry value chain to ensure product interoperability and growth of wireless market.

WCA (Wireless Communications Association International): A nonprofit trade association representing the wireless broadband industry.

WECA (Wireless Ethernet Compatibility Alliance): An organization that certifies interoperability of WiFi (IEEE 802.11) products and to promote WiFi as the global wireless LAN standard across all market segments.

WISPA (Wireless Internet Service Provider Association): A nonprofit organization and cooperative, formed to serve the interests of WISPs and ISPs worldwide.

WLANA (Wireless LAN Association): A nonprofit educational trade association, comprising the thought leaders and technology innovators in the local area wireless technology industry.

WCA (Wireless Communications Association): A nonprofit trade association, representing the wireless broadband industry, whose mission is to advance the interests of the industry's operators, equipment providers, and professional services firms.

WRA (WirelessReady Alliance): An alliance aimed at creating market awareness about the possibilities of wireless data technology.

APPENDIX

B

Useful URLs

Organization	Web Address

ACM Sigmobile
www.acm.org/sigmobile

Advanced Television Systems Committee
www.atsc.org

All Net Devices
http://devices.internet.com/

America's Network
www.americasnetwork.com

Anywhereyougo.com
www.anywhereyougo.com

Arbitron Radio Ratings/Research
www.arbitron.com

ATM Technology Forum
www.atmforum.com

Bank Insurance & Securities Assn.
www.bsanet.org

Bluetooth Consortium
www.bluetooth.com

Bluetooth Resource Center
www.palowireless.com/bluetooth/

Business 2.0 Magazine
www.business2.com

Cap Gemini Ernst & Young, US
www.us.cgey.com

Cap Gemini Ernst & Young, Worldwide
www.cgey.com

CGE&Y Center for Business Innovation
www.cbi.cgey.com

Cellular News
www.cellular-news.com

CDMA Development Group
www.cdg.org

CTIA (Cellular Telecommunications & Internet Association)
www.wow-com.com

Communication Systems Design
www.csdmag.com

eBiquity.org
www.ebiquity.org

European Telecommunications
www.etsi.org

Everything Wireless
http://web2.wireless.com/

Fierce Wireless
www.fiercewireless.com

GSM World
www.gsmworld.com

Handago.com
www.handango.com

Home RF/SWAP
www.homerf.org

IEEE
www.ieee.org

Internet Engineering Task Force
www.ietf.org

InternetNews.com
www.internetnews.com/wireless

Living Internet, The
www.livinginternet.com

Microsoft Research
http://research.microsoft.com

Microsoft Mobility
www.microsoft.com/windowsmobile

Microsoft Mobility Case Studies
www.microsoft.com/mobile/enterprise/casestudies/default.asp

Mobile Review
www.acm.org/sigmobile/MC2R

Mobile Data Association
www.mda-mobiledata.org

Mobile Info
www.mobileinfo.com

Motorola Bluetooth
www.motorola.com/bluetooth

National Association of Broadcasters
www.nab.org

National Emergency Number Assn.
www.nena9-1-1.org

Network World Fusion
www.nwfusion.com/topics/wireless.html

Open Mobile Alliance
www.wapforum.com

Pocket PC Magazine
www.pocketpcmag.com

PDA Buzz
www.pdabuzz.com

Red Herring Magazine
www.redherring.com

Strategic News Service
www.tapsns.com

SyncML Forum
www.syncml.org

TelecomWeb
www.telecomweb.com

Thinkmobile
http://thinkmobile.com

UMTS Forum
www.umts-forum.org

VoiceXML Forum
www.vxmlforum.com

WAP Sight
www.wapsight.com

Webopedia
www.webopedia.com

WiFi Alliance
www.wirelessethernet.org

Wired Magazine
www.wired.com

Wireless and WAP
www.itworks.be/WAP

Wireless Asia
www.telecomasia.net/telecomasia/

Wireless Design & Development
www.wirelessdesignmag.com

Wireless Developers Network
www.wirelessdevnet.com

Wireless News Factor
www.wirelessnewsfactor.com/

Wireless Systems Design
www.wsdmag.com

Wireless Week
www.wirelessweek.com

World Wide Web Consortium
www.w3c.org

Yahoo Wireless News
http://story.news.yahoo.com

APPENDIX C
Glossary

Acronym **Explanation**

3G International Telecommunications Union (ITC) specification for the third generation of mobile communications technology

AMPS Advanced mobile phone system

ARPU Average revenue per user

ASP Application service provider

B2B Business-to-business

B2C Business-to-consumer

B2E Business-to-employee

BSC Base station controller

CDMA Code division multiple access

CDPD Cellular data packet data

CHTML Compact HTML

CORBA Common Object Request Broker Architecture

CRM Customer relationship management

DES Data Encryption Standard

E911 Enhanced 911

ECMA European Computer Manufacturers Association

EDGE Enhanced Data Rates for Global Evolution

EMS Enhanced Message Service

ETSI European Telecommunications Standards Institute

FCC Federal Communications Commission

FoIP Fax over IP

G2C Government-to-citizen

GIF Graphics Interchange Format

GPRS General Packet Radio Service

GPS Global positioning system

GSM Global System for Mobile Communications

HDML Handheld Device Markup Language

HTML HyperText Markup Language

HTTP HyperText Transfer Protocol

HTTP-NG HTTP Next Generation

HTTPS HTTP Secure

IEEE Institute of Electrical and Electronics Engineers

IMT2000 International Mobile Telecommunications 2000

IP Internet Protocol

IPSec IP Security

Ipv6 IP version 6

ITU International Telecommunications Union

ISO International Organization for Standardization

ITRS International Technology Roadmap for Semiconductors

LAN Local area network

LDAP Lightweight Directory Access Protocol

MSC Mobile switching center

MSN Microsoft Network

NMT Nordic Mobile Telephone

OMC Operations and maintenance center

OS Operating system

OSI Open Systems Interconnection

PAN Personal area network

PCS Personal Communication Service

PDA Personal digital assistant

PDC Personal Digital Cellular

PIM Personal Information Management

PPP Point-to-Point Protocol

QA Quality assurance

QoS Quality of service

RF Radio frequency

RFID Radio Frequency ID (RFID)

ROI Return on investment

RTOS Real-Time Operating System

SDK Software Development Kit

SIM Subscriber Identification Module

SLA Service-level agreement

SLP Service Location Protocol

SMIL Synchronized Multimedia Integration Language

SMS Short Message Service

SNR Signal-to-noise ratio

SQL Structured Query Language

SSL Secure Socket Layer

SWAP Shared Wireless Application Protocol

SyncML Synchronization Markup Language

TACS Total Access Communications System

TCP/IP Transmission Control Protocol/Internet Protocol

TDMA Time division multiple access

TDOA Time difference of arrival

TETRA Terrestrial Trunked Radio

TIA Telecommunications Industry Association

TLS Transport Security Layer

TTS Text to speech

UDP User Datagram Protocol

UI User interface

UIML User Interface Markup Language

UMTS Universal Mobile Telecommunications System

URL Uniform Resource Locator

VLR Visitor Location Register

VML Vector Markup Language

VoIP Voice over IP

VXML Voice-Extensible Markup Language

W3C World Wide Web Consortium

WAE Wireless application environment

WAG Wireless Assisted GPS

WAN Wide area network

WAP Wireless Application Protocol

WASP Wireless application service provider

WATM Wireless Asynchronous Transfer Mode

WBMF Wireless Bitmap Format

WCDMA Wideband CDMA

WDP Wireless Datagram Protocol

WIEA Wireless Internet Enterprise Applications

WiFi Wireless Fidelity (802.11b WLAN)

WLAN Wireless local area network

WML Wireless Markup Language

WSP Wireless Session Protocol

WTA Wireless telephony applications

WTLS Wireless Transport Security Layer

WTP Wireless Transactional Protocol

XHTML Extensible HyperText Markup Language

XML Extensible Markup Language

XSL Extensible Style Language

Index

G

H

I

J-K

L

R

S